English for *Industry*

W. Büchel
R. Mattes
H. Mattes
M. Schäfer
Dr. W. Schäfer

8 rue Férou
75278 Paris cedex 06
www.editions-belin.com

BELIN

ISBN 2-7011-38

Foreword

The globalisation of markets for industrial products and services has made it essential for all those working in industry to be able to understand and express themselves in English, the language which has undoubtedly become the main vehicle of international communication.

The ability to use English is not only necessary for those people who deal with international sales and marketing but on other levels too. It concerns, among others, the person on the shop floor who has to understand instruction manuals, the technician who provides after-sales service to English-speaking users of his or her company's product, the Internet user who needs to understand technical specifications in an on-line catalaogue, the service engineer who is doing a training course with an English company...

English for Industry is therefore designed for all those who are studying English as part of a course leading to a technical qualification (e.g. the BTS specialities in mechanical or electrical engineering, electronics, metalwork...) as well as for people who need English in a professional capacity (technicians, trainers, line managers...). It may be used in the classroom or, with the aid of the "key", as a basis for self-study.

Each of the ten units is centred on one specific area of interest. These range from the basic tools of the trade ("**Information technology – Electronic communication**", "**Mechanical Engineering – Tools**"...) to wider topics such as "**Energy and the environment**" and "**Finding a job in Europe**". Within each unit, students learn the relevant technical vocabulary and expressions, and are placed in everyday situations that improve their communication skills – all the language skills are covered (both oral and written comprehension, and both oral and written production).

There are a wide range of activities, from controlled vocabulary work through to guided role play and project work. For those using the book in class, pair-work and groupwork activities are provided. A large number of recorded dialogues, indicated by the symbol on the right, give ample opportunity for oral comprehension and provide a springboard for further activities in oral and written production.

At the end of each unit, a **Grammar Revision** section provides practice in some areas of difficulty and, at the end of the book, a **Grammar Overview** deals with the major points of English grammar, providing a reference section for students who need to review specific points.

Also at the end of the book are the **Tapescripts** and two vocabulary lists, from English to French: **Unit vocabulary** and an **Alphabetical Wordlist**. Various tables, notably **Conversion tables**, complete the manual.

This student's book is complemented by an audio CD and a key which includes crosswords for use in class or at home.

Contents

Unit 1

The new job

Michel Arnaud is a young French engineer who passed his exams in mechanical engineering two years ago. After working as a manufacturing engineer in the development of new types of engines he decided to look for a new job on the Internet. After a short search he found the following job vacancies:

CAREERS AT POWER ENGINES / LEEDS **CURRENT OPPORTUNITIES**

Power Engines Ltd. is currently looking for highly motivated and skilled people to fill the current job vacancies listed below.
We would like to receive the CVs of anyone interested in a position within our company even if there do not appear to be any current opportunities within your particular field of specialization. Power Engines has a policy of keeping on file all CVs that we feel might meet our future recruitment needs, and when seeking to fill new positions we actively consult our files.
Should you wish to submit an application for one of the current positions at Power Engines, please ensure that you quote the appropriate file number.
We would like to thank all applicants in advance for their interest. Due to the anticipated large number of applicants we regret that we are able to contact only those candidates selected for an interview. For the convenience of the applicants, all interviews will be conducted at our London branch.
The following employment opportunities are currently available:

- 220-21G Electrical Engineer – Power
- 220-20G Network Systems Engineer
- 203-43G Test Engineer
- 203-01 Service Technician
- 200-09 Manager, Product Safety
- 114-219 Quality Engineer
- 114-06 Manufacturing Engineer
- 111-45G Research Engineer
- 22-27G Development Engineer
- 22-20G Sales Representative
- 11-23G Staff Accountant
- 11-03 Marketing Manager
- 04-01W Network Support Technician

home | products | corporate info | careers | site map

After carefully reading the brief outline of current vacancies above, Michel decided to look at two particular positions in more detail:

FILE NUMBER: 114-06 Manufacturing Engineer

Responsibilities:

Working as part of an international team, you will be involved in the development of fuel-cell components. The post is varied and offers the opportunity to be directly involved in all stages of manufacturing, from the initial research and development to the preparation for the market introduction of fuel-cell products.

Requirements:

Applicants must possess an Engineering degree and have a minimum of one year's experience in a manufacturing environment, preferably in a high-tech sector. We are looking for team players who thrive in a challenging environment. Knowledge of a second language, although not essential, would be desirable. If you wish to apply for this position, please do not forget to quote the above file number on all correspondence. Power Engines Ltd. will directly contact only those candidates they wish to interview.

Salary:

£30,000 per annum + company car

FILE NUMBER: 04-01W Network Support Technician

Responsibilities:

The successful candidate will be responsible for the management and maintenance of a computer network currently consisting of more than 400 PCs across a wide area network (WAN). Our network client base is expanding constantly and office network services have to be continually updated.

Requirements:

Candidates must have experience in the following areas:

2 or more years supporting Windows NT Server & Workstation; MS Mail/Exchange Server, installations and maintenance; experience in Wide Area Network (WAN) support.

Salary:

£35,000 per annum + fringe benefits

If you feel that you have the necessary skills and qualities for this position we would welcome your application. Please quote the file number on all documentation. Power Engines will contact those candidates to be interviewed. If you are unsuccessful, please accept our thanks for the interest you have shown in Power Engines Ltd.

back to current opportunities

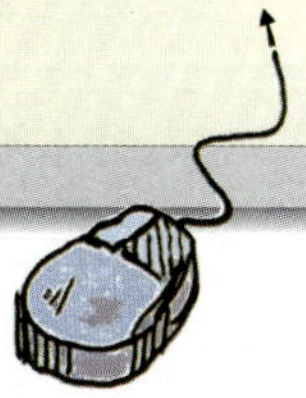

a Find synonyms in the advertisements for the following words.

1. candidates
2. job vacancies
3. advantageous
4. yearly
5. get in touch with
6. name, state
7. document
8. diverse
9. qualifications, preconditions
10. demanding
11. additional advantages
12. develop, do well

b Answer the following questions.

1. Which of the two jobs do you think Michel applied for? Give reasons for your choice.
2. Which job would interest you more? What is the reason for your choice?
3. Suppose you meet the requirements. What should you do next?
4. How would Power Engines Ltd. react to your application?

**c Describe the responsibilities and requirements for the two jobs.
Use the phrases in the box below.**

experience

right/suitable qualifications

good/better job prospects

potential for career development

to enjoy working in a team (teamwork)

better pay (salary)/ benefits

Example:
I would be more interested in applying for the position of manufacturing engineer because the job is varied and it involves teamwork.

Phrases: Giving your opinion

Phrases:	Explanations:
I think (that) ...	The most common way of giving your opinion.
I believe (that) ...	Used in formal speech or writing.
In my opinion ...	Mainly used in writing.
From my point of view ...	To express your personal view.
As far as I'm concerned ...	This emphasises your personal opinion – no matter what other people think.

d Look at the list of job vacancies on pages 7 & 8 and find the English words for the following:

> technicien de maintenance de réseau – ingénieur chercheur – ingénieur électricien – comptable – fabrication – candidature – actuel – entretien – avantages en nature – équipe – représentant – souhaitable – courrier

A Welcome to Leeds Airport!

A short time after his successful interview in London, Michel is on his way to his new job in Leeds. He has arrived at Leeds Airport, and after passing through customs and baggage reclaim he is now in the arrivals hall. He wants to get a taxi from the airport to his hotel in the city centre. He stops a porter and asks him the way to the taxi rank.

a Read the dialogue in pairs.

Michel: Excuse me, please. I was wondering if you could help me?
Porter: Yes, certainly. What can I do for you?
Michel : Could you tell me how to get to the taxi rank, please?
Porter: Well, let's see. You need to carry on along this corridor until you get to the newsagent's on your left. Turn left following the sign for city centre/buses. Go straight on for a few minutes until you come to the Tourist Information Desk on your right. Turn right here, walk past Tourist Information and you will come to some double doors. Go through the doors and turn right and there's the taxi rank.
Michel : Thanks a lot. So that's go straight on until I get to the newsagent's on the left. Turn left, go straight on again until I come to Tourist Information. Turn right there, go through the double doors and the taxi rank is on the right.
Porter: Yes, that's it.
Michel : Thanks again.
Porter: You're welcome.

b Imagine that you have just arrived at Leeds Airport. Working in pairs, choose a starting point and ask for and give directions to various places within the airport. Use the model dialogue on page 10 and the floor plan and the phrases below.

Example:

Excuse me, please. Could you tell me how to get to the customs hall?

– Yes, certainly. Go straight ahead …

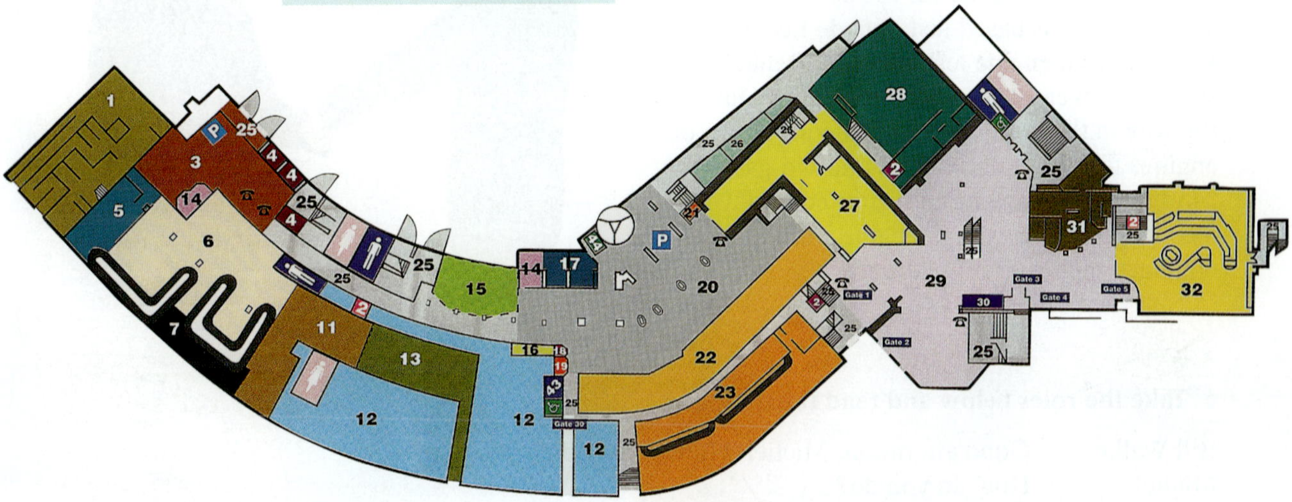

- 1. Domestic Arrivals Hall
- 4. Car Hire
- 5. Customs Hall
- 6. Baggage Hall
- 11. Arrivals Hall – Eire and Channel Islands
- 12. Airport Offices
- 13. International Arrivals Hall
- 14. Bureau de Change
- 15. Airport Shop
- 22. Check-in Desks
- 24. Main Entrance / Exit
- 27. Duty Free Shop
- 28. Central Screening Zone
- 29. International Departure Lounge
- 31. Yorkshire Hero Pub
- 32. White Rose Café

Phrases: Giving directions

Possible questions:

Excuse me, please. Could you tell me how to get to (the) …?

Is there a … nearby?

I was wondering if you could tell me the way to (the) …?

Possible responses:

Yes, certainly. Go straight ahead until you get to …

Yes, there is.

Turn right/left …

Pass the …

Go across …

… and the … is on your left/right.

B Arrival at Power Engines Ltd

After dropping his things off at his hotel Michel makes his way to Power Engines, where he is to meet Bill Walker, the Personnel Manager. Power Engines is a privately owned company which has been developing and manufacturing alternative fuel engines for the past 25 years. The company is based just outside Leeds. It has its own engine test facility and machine shop. For over 20 years, Power Engines has led the way in the development and conversion of engines to run on natural gas and bio-gases. Recently the company has also started to develop a completely new type of engine – the fuel-cell engine, which runs on hydrogen.

a Take the roles below and read the dialogue.

Bill Walker:	Good afternoon, Michel. How do you do?
Michel:	How do you do?
Bill:	Did you have a pleasant flight?
Michel:	Yes, thank you. Although there was a bit of a delay.
Bill:	Sorry to hear that. Please take a seat. Would you like something to drink? A cup of coffee or perhaps some tea?
Michel:	Tea would be fine, thank you.
Bill:	Milk and sugar?
Michel :	Just milk, please. Thank you.
Bill:	Right. Let me introduce you to your colleagues. This is Roberto Garcia from Spain ...
Michel :	Pleased to meet you.
Roberto:	Pleased to meet you, too.
Bill	... and Jerry Miller from the USA.
Jerry:	Nice to meet you, Oliver.
Michel :	Nice to meet you.
Bill:	... and this is Annette Mair from Austria. I believe you've met before?
Oliver:	Yes, we met at the interview. Nice to see you again, Annette.

After some small talk Bill Walker introduces Michel to Steven Hill.

Bill:	Ah, there's Steven. Michel, may I introduce you to your new boss? This is Steven Hill, who is in charge of production, engineering and quality control.
Steven:	How do you do? Pleased to meet you and welcome to Power Engines. I hope you'll enjoy working for us.
Michel:	I'm sure I will.
Bill:	Before Steven shows you around the various workshops, I'd like to give you a brief outline of our company ...

b Complete the following sentences with words from the text and the dialogue on page 12.

Power Engines has been developing and ① alternative ② for over 25 years. These engines run on ③ and ④ . One recent development is the fuel-cell engine, which runs on ⑤.
At Power Engines Ltd. Michel meets Mr Bill Walker, the personnel manager. He introduces him to his new ⑥. He also meets Steven Hill, who is in ⑦ production, engineering and ⑧.

INFO-BOX: Introducing people and saying goodbye

- **When introducing people you would normally use one of the following phrases:**

This is …
May I introduce you to …?
Have you met …?

- **The people who have been introduced then greet each other with the following phrases:**

'How do you do?'	Used when people meet each other for the first time. The response to this greeting would be 'How do you do?'.
'How are you?'	Used at the start of the conversation as a means of breaking the ice. When used in a formal context the speaker is not really asking how the person is. The person asking the question would normally only expect a standard reply such as 'Fine thanks (and you?)' or a variation such as 'I'm fine' or 'Fine, thanks'.
'Nice to meet you'.	Usually answered with 'Nice to meet you (, too)'. Another possible greeting is 'Pleased to meet you'.

- **Saying goodbye:**

'See you later/on Monday'.	(informal) Normally used with good friends who you will see again.
'It was nice meeting you'.	(formal) Used with people that you have met for the first time.

Note: First names are commonly used in England and the USA.
Ms ... is used for both married and unmarried women.

c Role play

Three students form a group, one takes the leading role and introduces Student A to Student B. Use the phrases you have learned above and try to fit them into the situations on page 14. Change over roles.

Example:	Introducing person:	"Mr Harris, may I introduce you to Mr Walker from Power Engines?"
	Mr Harris:	"Nice to meet you, Mr Walker."
	Mr Walker:	"Nice to meet you, Mr Harris. Did you have a good journey?"
	Mr Harris:	"Yes, thank you. Although there was a lot of traffic."

Student A	Student B	Possible questions / answers
Mr Harris from Techno Plast in Sheffield (as given in example)	Mr Walker, personnel manager, Power Engines	good journey? / yes, but a lot of traffic
Dave Rosevere from Glasgow	Steven Hill, production manager, Power Engines	train? / not too bad / 10 minutes' delay
Mr Delaney, applicant for a job at Power Engines Ltd	Jane Owttrim, purchasing / stores, Power Engines	problems finding the company? / no, followed the directions given
Ms Jensen, a student from Lille	Geoff Harris, Accounting, Power Engines	weather in Lille? / rain / a bit cold
Helen Miller from the Manchester Observer (newspaper)	Anna Carter, public relations manager, Power Engines	apologises for being late / no problem

d Introduce yourself and continue the conversation as follows. Use the phrases in the box below.

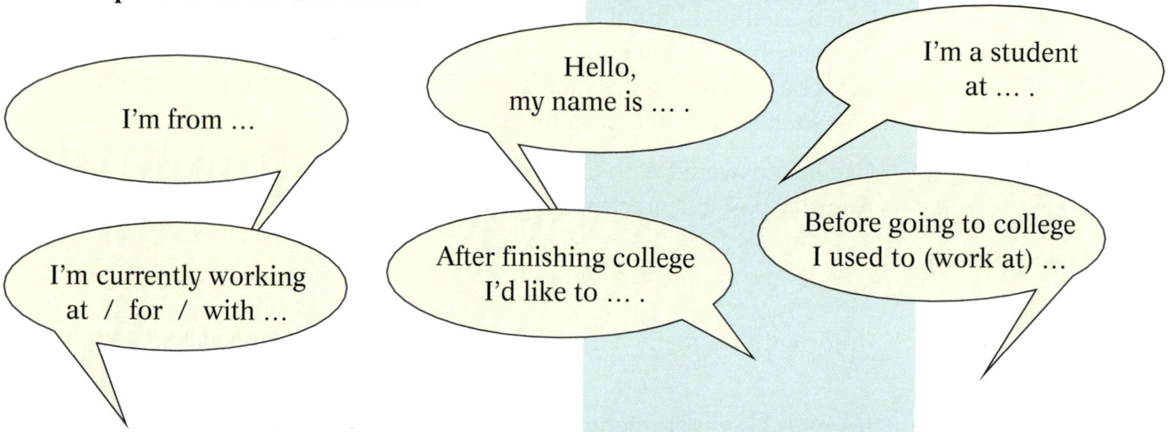

Phrases: Welcoming a foreign visitor

Host	Foreign colleague / visitor
First meeting	
Welcome to …	Thank you. It's nice to be here.
How do you do? (formal)	How do you do? (formal)
Pleased / Nice to meet you.	Pleased / Nice to meet you, too.
I'm … / My name is …	I'm …
Please call me …	You can call me …
Second / third … meeting	
How are you? (informal)	Fine, thanks. And you? (informal) Not too bad.
Hello, … / Hi, … (informal)	Hello / Hi (informal)

Introducing someone

This is …	Nice to meet you.
I'd like you to meet …	
Have you met …?	

The journey / flight

Did you have a pleasant / comfortable flight / journey / trip?	Yes, thank you.
How was the flight / journey / trip?	It was very comfortable.
What was the flight like?	The service was excellent.
How long did it take from … to …?	… hours.

General conversation (small talk)

Have you been to … before?	No, I've never been to (this part of) … before.
What is your country like?	It's quite different from … / It's very similar to …
Would you like to meet / see / visit ...?	Yes, that would be fine.
Have you ever tried …?	No, not yet, but I would love to try one / some.
We've arranged a … for you.	Thank you, that's very kind of you.
Tonight, we're having a party / reception / celebration.	Yes, I'd love to come.
Would you like to join us?	Thank you for the invitation, but I'm afraid I can't. I'm meeting …
Are you married?	Yes, I am. / No, I'm single / divorced.

The work

What do you do?	I'm in electronics / computers …
Who do you work for?	I work for / at / with …
	I'm self-employed.
What sort of work will I be doing?	Your duties will be ...
Where is the … department?	The … department is on the … floor.
Who's in charge of / responsible for …?	Mr / Mrs / Ms …

e Role play: How would you reply to the following? Take turns with your partner.

1. How do you do?
2. How are you?
3. Nice to meet you.
4. I'd like you to meet Mr Cole, our marketing manager.
5. Did you have a good flight?
6. I'm sorry for being late.

C Introducing the company

After the meeting Peter Thompson from research and development explains the company structure to Michel Arnaud with the following organigram.

Who is who? Organisation Tree of Power Engines Limited

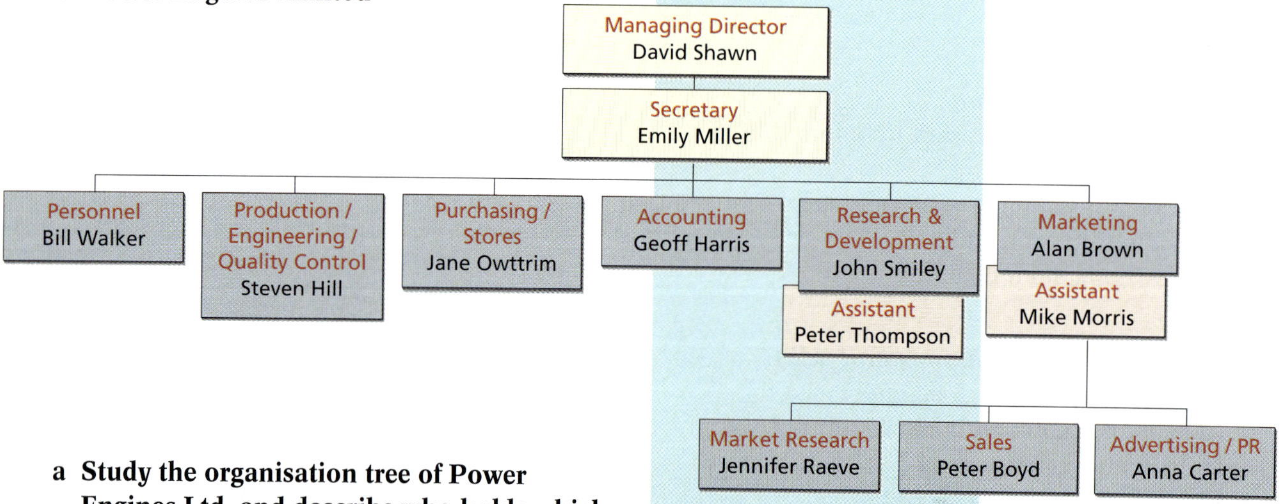

a Study the organisation tree of Power Engines Ltd. and describe who holds which position within the company. Form sentences as in the example.

Example:
David Shawn is the managing director.

b Look at the organigram above and match the following tasks with the departments.

- Make offers.
- Order new goods for production.
- Check quality of all products.
- Advise customers.
- Pay employees' salaries.
- Place advertisements in special interest magazines.
- Do surveys and collect statistics.
- Store raw materials and finished goods.
- Hire new employees.
- Plan and design new products.

c Look at the plan of the company's premises. Then read the texts below and find the destinations described.

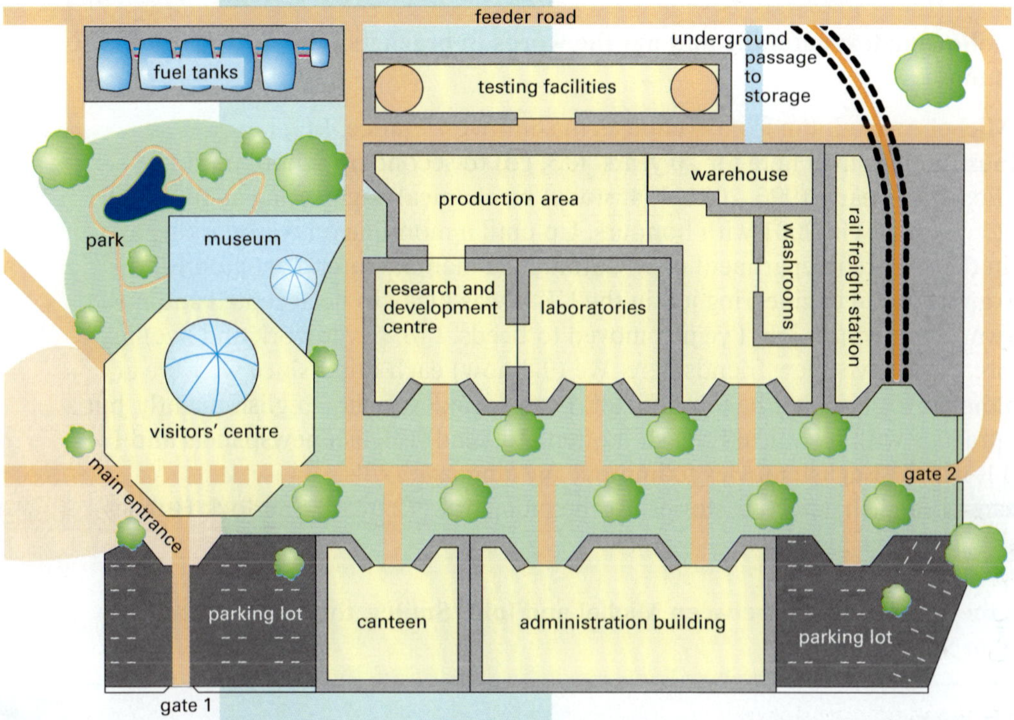

1. Michel Arnaud drives in at the main entrance to the premises of Power Engines Ltd., but he finds a parking space only in the lot near gate 2. He leaves the lot and walks left, past the administration building and the canteen, and directly ahead of him is the octagonal visitors' centre. He signs in there and follows the signs to the building attached to it, the ?, where he plans to look around and get an idea of the history of the company before meeting the personnel manager, Bill Walker, for a tour of the premises.

2. The tour of the company begins in the research and development (R&D) centre, right next to the museum. After John Smiley has described the projects that are being planned, Michel is taken next door to the laboratories, where his future colleagues Roberto, Jerry and Annette show him what they are working on. When they leave the labs they turn left, and left again, and enter the ?.

3. They walk out of the production area and turn right. They walk past the washrooms on the left and leave the building. They turn right and pass the administration building on their left. They finally arrive at the ?, opposite R&D.

d After lunch Michel visits the testing facilities where his tour ends. He must return to the parking lot at gate 2. He asks for directions, being somewhat confused after having seen so much in one day. **How can he get there without passing the rail freight station, which is very busy and crowded at this time of the afternoon?**

e Listen to the company tour description and take notes.

D Grammar revision: **Tenses** (Present simple, present continuous, past tense or present perfect?)

See Grammar Overview in the appendix.

a Study the following text carefully and put the words in brackets into the correct tense.

My name is Michel Arnaud. I'm from Toulouse, in southwest France. I (1. be) born in Toulouse and (2. live) here for 20 years. I (3. go) to secondary school until I was 18. For the next 5 years, 1996–2001, I (4. study) mechanical engineering at Toulouse University. I (5. graduate) in 2001 with honours. Up until a month ago I was working as an engineer in the research department of an automobile manufacturer. Last month I (6. decide) to apply for an engineering job in the UK. I was lucky to be accepted and now I (7. look forward) to my new job. I've just moved to Leeds. I (8. never work and live) in England before, but I have a few friends here. We (9. know) each other since we were at university so the move shouldn't be too difficult. I can speak everyday English fluently, but I need some practice with technical English. I often (10. read) English newspapers and I sometimes (11. watch) English movies to improve my language skills.
And I like reading English detective stories. At the moment I (12. read) Sherlock Holmes stories.

b Complete the short dialogue between Michel and John Smiley, the head of R&D. Use 'since' or 'for'.

INFO-BOX: ***for*** **and** ***since*** **= depuis**

since: relates back to precise moments in time or dates (e.g. since 1995; since Christmas; since he left school), and is used for asking questions about dates (Since when ...? *Depuis quand ... ?*)

for: relates to periods of time (for five minutes; for 3 days; for 2 years)

Michel: 'How long have you been the director of research and development, if you don't mind me asking?'

John: 'Well, I've been with Power Engines ① 20 years now, ② leaving school. The first couple of years I learned about the business and then worked my way up in the company. I was in purchasing ③ a few years and then moved into production. I got a bit of experience and was soon promoted. Two years ago I was promoted to the position of director. ④ then I've been working on future projects, that is projects for the next 5 years, well into the millennium.
You said that you worked in the R&D department in your last job.
What made you decide to go into research?

Michel: I've been interested in engineering ever ⑤ I went to a school workshop when I was about eight. I went to university to study mechanical engineering and after graduating I found a job as an engineer, but it was only ⑥ a 2-year short-term contract."

Unit 2

Office work

a Match the following pieces of office equipment and stationery with the pictures above.

stapler – hole punch – envelope – biros – tray – scissors – paper clips – pencil sharpener – clock – notepad

b What else can you see in the picture above?

A Office routine

Before Michel starts work as an engineer at Power Engines Ltd. he wants to become familiar with the office routines in his new company, so he is going to spend the first week in the purchasing department. There he will learn about Power Engines' suppliers, the costs of the materials used and the way people in modern companies communicate with each other. He will spend a week in Jane Owttrim's office, to find out who does what and where.

a Read the dialogue in pairs.

Jane: Michel, have you seen Mike today?
I asked him to copy the offer from Distel Electronics for our new control panels. And he should also send some faxes and write some letters.

Michel: No, I haven't seen him. But I could make the copies for you. Just tell me how many and who the copies are for. And later I could work on the letters and faxes.

Jane: Oliver, are you sure you don't mind? It's really nice of you to offer. I think four copies should be enough; let me think, one for Steven Hill, one for his assistant, one for Alan Brown and one I should keep in my files, so I know where to look when they call us. The photocopier is just down the hall.

Michel: OK.

Jane: And when you are done, please put the copies in their pigeonholes, just outside my door. Their secretaries will pick them up when they walk by.

b Michel takes the offer, consisting of six pages, and goes to the photocopier.
Put the following sentences in the correct order.

1. Afterwards he sorts the copies into four piles.
2. He puts each page face down on the glass screen.
3. He staples the copies together with a stapler.
4. He presses the number four on the controls.
5. He puts the copies into the pigeonholes.
6. He presses the print button.

c Match the words from the following list with the numbers in the sentences on the next page.

communicate – suppliers – purchasing – familiar – make – photocopier

1. The department of a company which is responsible for buying raw materials and goods is called the ① department.
2. When you have good knowledge about a certain matter you are ② with it.
3. Companies who provide other companies with material are called ③.
4. Modern technology has made it easier and faster for people to ④ with each other.
5. With a ⑤ you can duplicate all sorts of printed and hand-written matter, drawings, plans, maps, even photographs.
6. Photocopiers are the same all over the world, so it is easy to ⑥ copies.

d Answer the following questions in your own words.

1. What will Michel learn during his first week?
2. What does Michel have to do with the offer from Distel Electronics?
3. What does Michel do after he has made the copies?

e Give a short report to the class:

Describe the functions of the photocopier your company / your school works with. Is it a black-and-white photocopier, or can it also copy coloured originals? Can it make copies on both sides of a sheet of paper? Can it sort the copies? What paper formats does it use? How fast is it? Does it produce good-quality copies? How much does a single copy cost?

f When Michel returns, he finds two folders on his desk.
Read the following dialogue in pairs. Then give a short summary of it in French.

Jane: The red folder is the one with letters and faxes still to be worked on. Would you please send this fax to Distel Electronics? It's very important; the number was engaged all morning. If they don't get it by Monday, they'll think we're not interested in their offer any more. Mr Hill told me to accept it, you know. Maybe you could also type the enquiry to Jackson & Green in Liverpool about the oil pumps. The other folder is just stuff that has to be filed. The letters are all marked on the top right corner concerning where they have to go.

Michel: Yes, I see. This one says "Incoming Mail", another "Outgoing Mail". What do I have next? Ah, here, "Enquiries", "Offers", and "Orders"; here is another one, "Contracts" and one marked in orange. Do these letters go into the folder named urgent? Could you possibly help me with them?

Jane: Sorry, not right now Michel, would you just try the fax again? The filing can wait until later.

So Michel takes the letter addressed to Distel Electronics, checks the fax number on the address and puts it into the slot of the fax machine. He dials the 0 that is necessary to get an outside line, then he dials the area code and the number. There is no busy signal this time, and his fax goes through within half a minute.

g Explain your personal filing system at home:

private letters – insurance – monthly expenses (rent, electricity, water) – work – school – documents

h Practise polite questions and answers with your partner. Use the suggestions in the box below.

Example:
pass me the folder / positive answer.
Could / Would you please pass me the folder? – *Yes, certainly / of course.*

1. open the window / positive answer
2. repeat the last question / positive answer
3. help me write a letter /polite no
4. translate a letter into French / polite no – maybe later
5. explain the fax machine / positive answer, but a little later
6. make some copies for me / positive answer

Phrases: Polite questions and answers

Questions:	**Answers:**
Would you please / Could you please ... I wonder if you could help me with ... Would / Could you do me a favour ...	Yes, certainly; yes, of course; sure thing; yes, of course / I'm sorry, I can't right now. I'd love to, but I'm busy myself at the moment. Oh, absolutely / I'll just finish what I'm doing and then I'll be right with you.
Would you mind + -ing e.g. "Would you mind copying a few letters for me?"	No, not at all.

i Michel's first week is almost over; he has already seen much of what goes on in a company. He knows a lot of the people, and Jane has been especially helpful. It is Friday afternoon, and there is still a lot to do.
Jane: "Michel, would you do me a favour, please? I would like to leave a little earlier today. Could you manage alone until four o'clock?"
Michel: "Well, certainly, I know my way round now."
What do / did you usually do on your Friday afternoons at work? Give a short account.

I always ...
I usually ...
I often ...
I sometimes ...
I occasionally ...
I rarely ...
I never ...

do the filing / tidy my desk / write letters and faxes /
phone customers /deal with complaints / write invoices ...

B Telephoning

Michel is tidying his desk and filing letters and notes. It is almost 3:15 p.m when the telephone suddenly rings. He has his notepad and pencil. ready and is prepared to take the call.

a Listen to the telephone conversation and find out what it is about. What is the main point of the call?

b Have a notepad ready. Listen to the telephone conversation again and take notes to answer the following questions.
(The time of the call: 3:15 p.m., date: Friday, 2 June ...)

1. Who is calling from where?
2. Who do they want to speak to?
3. What is the subject of the call?
4. What do the caller and Michel agree on?
5. Who should Mr Hill contact on Tuesday, in case Mr Slate isn't in?

c Take your notes and turn them into an urgent message for Mr Hill.

d Use the spelling table below and spell your name / your neighbour's name, the names of your school, company etc.

Spelling Table

	Great Britain	***USA***		***Great Britain***	***USA***
A	Andrew	Abel	**N**	Nellie	Nan
B	Benjamin	Baker	**O**	Oliver	Oboe
C	Charlie	Charlie	**P**	Peter	Peter
D	David	Dog	**Q**	Queenie	Queen
E	Edward	Easy	**R**	Robert	Roger
F	Frederick	Fox	**S**	Sugar	Sugar
G	George	George	**T**	Tommy	Tare
H	Harry	How	**U**	Uncle	Uncle
I	Isaac	Item	**V**	Victor	Victor
J	Jack	Jig	**W**	William	William
K	King	King	**X**	Xmas	X
L	Lucy	Love	**Y**	Yellow	Yoke
M	Mary	Mike	**Z**	Zebra	Zebra

e Telephone conversation: Give the missing answers for the conversation between Mr Hill and Michel on Monday morning or write down the answers on a separate sheet of paper.

Mr Hill: Hello ….
Michel: Good morning, Mr Hill. This is Michel Arnaud speaking.
Mr Hill: Did you have a nice weekend, Michel?
Michel: ….
Mr Hill: Fine, what's on your mind, Michel?
Michel: ….
Mr Hill: A new offer, and cheaper you say? How much cheaper?
Michel: ….
Mr Hill: And how can they suddenly be cheaper?
Michel: ….
Mr. Hill: How much time do we have to decide on the new offer?
Michel: ….
Mr Hill: Well thank you, Michel, that certainly is good news for a Monday morning.

Phrases: Telephoning

In England and America people normally answer the telephone at home not with their name but with *Hello*, or with their own number.
It is the caller who gives his or her name first and states who he or she wishes to speak to. However, if you telephone with a business or an office the call will be answered with the company name.

What to say if you are making a call:

- My name is
- I'd like to speak to / Could / May I speak to / Would it be possible to speak to …
- I'm calling from (France / from Manchester Airport) …
- I'm phoning about … (our order / your invoice / your enquiry)
- Could you give me the number / any information on / about …
- Would you please call me back at this number …?
- Can you put me through to …?
- I'm sorry, I didn't understand …
- I'm sorry, I didn't catch that.
- Could you spell the name please?
- Could you speak up a bit please? The connection is bad / we have a bad connection.
- Sorry, I didn't hear you / didn't understand it correctly; could you repeat it please?
- I'm not sure if I understood correctly. May I repeat it back to you?

Phrases you may hear as a caller:

- Sorry, wrong number.
- Please leave a message on my / our answering machine.
- Please hold the line (used by automatic machines).

- What's your name again? / Who did you say was calling?
- Who was it you wanted to speak to?
- Can we call you back?
- Hold on [formal] / Hang on [informal].
- I'm afraid the line is busy.
- Can I take a message?
- Would you like to leave a message?
- I'm afraid Mr/Ms … is not in / is in a meeting just now / won't be back before Friday / is on holiday.
- I'll have him return your call.

Phrases you may need if you receive a call:

- Michel Arnaud speaking.
- How can I help you / be of assistance? / What can I do for you?
- Hold on, I'll have a look /check our files.
- May I call you back in a few minutes / this afternoon / tomorrow?
- I'll get back to you as soon as possible / this afternoon / when I've spoken to Mr / Ms …
- I'll get that out in today's mail.

f Calling directory enquiries: Listen to the CD and write down the numbers that directory enquiries give to the callers. Note that the number 0 is usually pronounced as "oh".

INFO-BOX: Directory enquiries

directory enquiry – Renseignements

telephone directory – annuaire

yellow pages – pages jaunes

country code – code pays

area code – code région

extension / extension number – poste / numéro de poste

Could I have / Could you give me the number of …

I would like to have the number of / for …

The line is engaged. (UK)
The line is busy. (US)
} Le numéro est occupé.

g Ask your neighbour to give you the number of his company; ask about numbers of institutions, your school, the town hall. It might be helpful to use your local telephone directory / the telephone directory of your company / your school for this role play.

h **Read the numbers:**

Example:

2.54 – "two point five four"

2,278 – "two thousand two hundred seventy-eight"

1. 3.14
2. 12.96
3. 6,197
4. 2,570,000
5. $13,800
6. 198,456
7. From the east coast to the west coast of the USA it is about 3,000 miles.
8. The Saturn V rocket carried over 2,000 tons of fuel.

INFO-BOX: Reading numbers

You do not pronounce the comma in numbers, but rather the decimal point.

Note: In English you have a comma where French has a full stop. The full stop in English is used with decimal numbers, the comma is used just as a means of orientation in longer numbers to separate the numbers into three-digit groups.

For example, you can buy a used car for **£3,999**, a meal in a restaurant for **£8.50 ("eight pounds fifty")**.

i Internet and e-mail addresses have to be handled with care. Even one little mistake in the punctuation will prevent you from finding the Website or will keep your e-mail from being delivered. By the way, "http" stands for "hypertext terminal protocol", and "www" for world wide web.

Listen to the examples on the CD:

1. **http://www.ft.com** (http colon double slash www dot ft dot com)
2. **http://www.irish-times.ie** (http colon double slash www dot irish hyphen times dot ie)
3. **http://www.city.net/countries/united_kingdom/england/london** (http colon double slash www dot city dot net slash countries slash united underline kingdom slash england slash london)
4. **http://www.virtual-london.co.uk/ghosties.html** (http colon double slash www dot virtual hyphen london dot co dot uk slash ghosties html)
5. **Site.Feedback@ft.com** (Site dot Feedback at ft dot com)
6. **info@phantasy.de** (info at phantasy dot de)

j **Now read the following addresses.**

1. http://www.yahoo.com
2. http://www.thelondon.net.co.uk
3. http://www.infoseek.com
4. http://www.altavista.digital.com
5. http://www.city.net/countries/united_states/new_york
6. http://www.number-10.gov.uk./tour/index.html
7. info@zgb-price.edu

k It is very important for Michel to know the different measures and weights and their French equivalents. Whereas most Western countries use the metric system, the United States and, to some degree, the United Kingdom, still use imperial measures.
For example, 1 ounce equals [avoirdupois weight] 28.3495 grams; for jewellers [troy weight] or pharmacists [apothecaries' weight] 1 ounce equals 31.1035 grams.
Note the differences between British and American weights and measures in the boxes below.

Look at the following tables of weights and measures, listen to the CD, and write down the measurements given.

Linear measure

1 inch = 2.54 centimetres (You read: 1 inch equals 2 point 54 cm) or in short form:
1 in = 2.54 cm
12 inches = 1 foot (You read: 12 inches equal 1 foot) or in short form: 12 in = 1 ft
1 foot = 0.3048 metre or in short form:
1 ft = 0.3048 m
3 feet = 1 yard, or in short form: 3 ft = 1 yd
1 yard = 0.9144 metre; or in short form:
1 yd = 0.9144 m
1 mile = 1.6093 kilometres; or in short form:
1 ml (in American English mi) = 1.6093 km

Square measure

1 square inch = 6.452 square centimetres
1 square foot = 929.03 square centimetres
1 square yard = 0.8361 square metre
1 acre = 0.4047 hectare

Weight

1 ounce (oz) = 28.350 grams
1 pound (lb.) = 16 oz = 0.454 kilograms
14 pounds = 1 stone = 6.35 kilograms
1 hundredweight (cwt) = 50.802 kilograms (n.b. long hundredweight = 50.802 kilograms, short hundredweight = 45.359 kilograms
1 short ton (t) = 1.016 metric tons
1 long ton = 0.907 metric tons

Cubic measure

1 cubic inch = 16.387 cubic centimetres
1 cubic foot = 0.0283 cubic metre
1 cubic yard = 0.746 cubic metre

Liquid measure, United States

1 fluid ounce = 29.573 millilitres
1 pint = 16 fluid ounces = 0.4732 litre
1 quart = 2 pints = 0.946 litre
1 gallon = 4 quarts = 3.785 litres

Liquid measure, Great Britain

1 fluid ounce (fl oz) = 28.412 millilitres
1 pint (pt)= 20 fl oz = 0.568 litres
1 quart (qt)= 2 pints = 1.136 litres
1 gallon (gal)= 4 quarts = 4.546 litres

C Letter writing

a Match the different types of written correspondence below with the following terms:

Business letter   Memo

Fax   E-mail

Dear Sir or Madam

Five years ago we purchased 30 control panels (product no. D-500) for our centre lathes. We and our clients were very satisfied with the results, and as we have now developed a slightly different version we would like to receive an up-to-date offer from you for another 30 panels (machine specifications are attached).

We would require delivery by 15 November at the latest, so please consider this condition when you send your offer. We look forward to receiving your offer as soon as possible.

Yours faithfully
Jane Owttrim,
Purchasing Department

①

Power Engines Ltd.
Combustion Drive
Leeds LSZ 5SE

Jackson & Green
156 Albert Docks
Liverpool 15P 5SN

18 May 200...

Dear Mr Green

Oil pumps

In response to your offer, we would like to order 100 oil pumps, series N-200, at the price and terms of payment quoted in your letter of 10 May. Regarding delivery, we definitely need the first 50 pumps by the end of October, but delivery of the remaining 50 just after the new year is acceptable.

Please give us at least one week's notice of delivery in October. We are looking forward to doing business with you on a regular basis.

Yours sincerely
Jane Owttrim

Jane Owttrim
Purchasing Department

②

To: Jane
From: Michel
28 May 200...

Have sent a fax to Distel Electronics saying you will accept offer next week.

Michel

③

Date: 28 May 200...
To: Distel Electronics, Manchester
Fax: 0161 876 50 01

From: Power Engines Ltd., Leeds
Fax: 0113 768 40 32

Dear Mr Johnson

We have received your offer for 30 control panels and will send a letter of acceptance at the beginning of next week, when the purchasing manager has returned from London.
We appreciate your prompt reply to our enquiry.

Yours sincerely

Michel Arnaud

Michel Arnaud, for Jane Owttrim
Purchasing Department

④

PE - Power Engines Ltd

Combustion Drive
Leeds

Your ref.:
Our ref.: ①

20 May ... ②

Matthews and Sons
Queen of Scots Ave.
Glasgow ③

Dear Sir or Madam ④

<u>Mikrolas 50 /article in Engineering Today, April issue</u> ⑤

In the latest issue of *Engineering Today* we read that you have developed laser equipment for measuring distances in the µ-range called *Mikrolas 50*. The article did not mention a price. As you may know, we develop new engines. Since the reduction of abrasion of moving parts in engines with high rpm is of tremendous importance to us, we are very interested in the applications of your equipment. Please send us detailed descriptions and quote your most favourable price as well as terms of payment and delivery. ⑥

We look forward to hearing from you as soon as possible.

Yours faithfully ⑦

Jane Owttrim

Jane Owttrim

Encl.: Copy of an article ⑧

b Match the numbers from the sample business letter shown above with the following terms:

complimentary close – subject matter – enclosure – salutation – reference –
inside address – body of the letter – date

INFO-BOX: Business letter – General information

Business letters represent your company, so they should be neatly typed or printed on your company paper. You should always write in a polite tone and the language you use should be clear – misunderstandings are a nuisance to everybody and they cost time. Keep your sentences short. Use new paragraphs to introduce new subjects.

Form and layout of a business letter

1. **Letterhead** (name, address of your company, telephone, fax, e-mail address)
2. **Inside address** "FAO" (For the attention of), "Attention" or "Attn." is added when the letter is addressed to a particular person in the company, e.g.:
   P.E. Power Engines Ltd.
   Attn. Michel Arnaud
   Combustion Drive

3. **Reference** (used as soon as there is an existing correspondence to refer to)
4. **Date**
   Note: There are various ways of giving the date in English:
   a. 2 June (2.6.) 2004 (you read: "the second of June 2004" – in GB
   b. June 2 (6.2.), 2004 (you read "June the second, 2004") – in the USA
   c. 2nd June 2004 – mainly in personal letters.
   To avoid confusion always write the name of the month as a word, never as a number.
5. **Salutation:** If your addressee is a single person ("Mr Godwin" or "Ms Wollstonecraft") address them as follows: "Dear Mr Godwin" or "Dear Ms Wollstonecraft" ("Dear Sir" or "Dear Madam" if you do not know their name). If you have been introduced to the person and are on a first name basis (which is common in English-speaking countries), use first names: "Dear William" or "Dear Mary".
   If you don't know any person or name write "Dear Sirs" (rather old-fashioned but still in use) or "Dear Sir or Madam". In the USA the salutation "Gentlemen:" is still in use.
   Notice the difference in punctuation: after "Dear Sirs", "Dear Sir", "Dear Madam", "Dear Sir or Madam" and "Dear Mary" use a comma or – more modern – no punctuation at all. "Gentlemen:", however, is followed by a colon (:).
6. **Subject line** Informs the reader quickly about the subject matter of a letter. It is usually placed after the salutation, but sometimes above it (as in the French style).
7. **Body of the letter** The first word is always capitalised, despite a possible comma after the salutation. Regardless of the different kinds of business letters, there is one basic structure common to all types: **introduction, main part and close.**
8. **Complimentary close** If the salutation reads "Dear Sirs", "Dear Sir"or "Dear Sir or Madam", close with "Yours faithfully". If you used the last name (Dear Mr Godwin) close with "Yours sincerely". If you are on a first-name basis: "With kind regards", "With best wishes" or just "Kind regards" or "Best wishes".
   In the USA, letters are closed with "Yours truly", "Yours very truly", "Sincerely yours" or "Very truly yours".
9. **Name and signature** The name is usually typed under the signature.
10. **Enclosure(s)** Abbreviated as "encl." and listed at the end of the letter.

c Write a letter of enquiry to Jackson & Green, 156 Albert Docks, Liverpool. Use Jane's notes on the right and the phrases below.

special offer for oil pumps in the June edition of **Technical Equipment**/would like to receive an offer for 100 pumps to be built into our engines. Quote your most favourable prices, terms of payment and delivery

Phrases: Business letter

introduction We see from your advertisement in the magazine … that
We found your address in the magazine …
We have been informed that …
We learned from our business partner / supplier that you …
The Chamber of Commerce has informed us that …
We saw your products while visiting the Cebit in Hanover …
We would like to revive our former business relations …
We would like to add your products to our product range …

subject of the letter Please send us your offer for …
Please let us have your prices for …
Please send us your catalogue and latest price list.
What are your terms of payment and delivery?
Please quote your most favourable prices / terms of payment / terms of delivery.
Please send us your offer for …, quoting your most favourable terms of delivery / payment.
What are the charges for …?
We would like to know your current rates for …
What would be your earliest possible delivery date?
You might want to take into consideration that we are bulk buyers.
Please note that we intend to buy large quantities if our first order meets with our customers' approval.

close We look forward to hearing from you soon.
Please send us your offer by return of post.
We would be pleased to hear from you soon.

d Complete the following part of a business letter using words from the box above:

We have been informed by the Chamber of ① that you are one of the leading manufacturers of lubricating oils. We are presently looking for a new ② . We use approximately 150 gal a year. Please send us an ③ for 10 gal. If the quality meets our standards we would like to ④ large quantities. Please quote your most ⑤ terms of ⑥ and ⑦ . We look forward to ⑧ from you soon.

e You have heard about a computer antivirus program developed in the USA by Virussystems, PO Box 4969, Trevose, PA 19053, and you want details about the product and the licensing costs. Your network has 40 users and you are interested in a licence for either one or two years. You also want to know if the program works for Internet and e-mail viruses. Write a letter using the correct layout.

D Grammar revision: Adjectives / Adverbs / Prepositions

See Grammar Overview in the appendix.

I Comparison of adjectives

- **Compare the following means of office communication. Use the adjectives in brackets.**

1. A photocopier copies a letter much (fast) than you can by hand.
2. Faxing a one-page letter is usually (cheap) than sending it by regular mail.
3. Sending a letter by regular mail is definitely (slow) than sending it by e-mail.
4. A laser printer is much (expensive) than a matrix printer.
5. The (expensive) printers do not always offer the (good) results.
6. Many people find it (easy) to make telephone calls than to write letters.
7. The (good) way to do it is to send him an e-mail.
8. This is the (bad) print quality I've ever seen.

II Adjective or adverb?

- **Use the correct form of the words in brackets.**

After ① a week Michel is quite ② with the office routine of the purchasing department. (near, familiar)
He can handle faxes and incoming letters ③ ④ now. (real, good)
But he has to work very ⑤, and often he goes home ⑥. (hard, late)
Jane has become a bit impatient with his many questions ⑦, but she ⑧ lets him notice it. (late, hard)
With technical equipment Michel has proven to be a ⑨ learner; he can handle e-mail ⑩ ⑪. (fast, remarkable, good)
He ⑫ enjoys working with his new colleagues. (obvious)

Note the different meanings:

hardly	– ne … guère	hard (adjective and adverb)	–	dur, difficile
nearly	– presque	near (adjective and adverb)	–	proche, près
lately	– récemment	late (adjective and adverb)	–	tard, en retard
		fast (adjective and adverb)	–	rapide, vite

III Prepositions

- **Put in the correct preposition** (by, on, to, at, of, with)

1. Peter is familiar … a lot … people.
2. Our next meeting is … Friday afternoon.
3. Everybody looks forward … a nice weekend.
4. Many working people have a family waiting for them … home.
5. Can you manage to get the job done … three o'clock?

Unit 3

Information technologies – Electronic communication

a Name the components of the computer workstation above.

b Which of the components are

1. input devices?
2. output devices?
3. storage devices?

c Match the words with the definitions. Form complete sentences.

1. Peripherals	a. Using different types of technologies such as audio and video in computing.
2. Laptop	b. A set of keys that operate a computer.
3. Multimedia	c. An input device for computer games.
4. Keyboard	d. A device used for making engineering drawings.
5. Joystick	e. An input device to move the cursor on the screen or to select from a menu.
6. Laser printer	f. A small portable computer.
7. Scanner	g. A storage device.
8. Mouse	h. A device that uses photoelectric cells to read texts or photos into a computer where these data can be manipulated and stored .
9. Hard disk	i. Any device connected to the central processing unit (CPU).
10. Central Processing Unit (CPU)	j. The place where data is temporarily stored in a computer.
11. Memory	k. Controls the basic functions of a computer. It consists of the control unit, the arithmetic and logic unit and the input/output unit.
12. Plotter	l. A device linked to a computer that produces good-quality printed material by means of a laser beam.

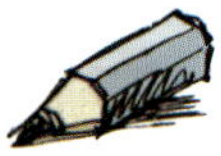

d Make a list of what computers can do.

Computers control ...

Computers speed up ...

Computers facilitate ...

Computers can help you when you need ...

communication

machines

paperwork

up-to-the-minute information

A Hardware

- **Give a definition of hardware by putting the following words in the correct order.**

1. such as / hardware / the physical components / are called / of a computer / the hard drive / the monitor / and

A1 The keyboard – the most important input device

a Match the names of the keys in the box with the descriptions of their functions below, and give their French equivalents.

- ESC
- Caps lock
- Shift
- Ctrl and Alt
- Space bar
- Enter
- Backspace
- Insert
- Delete
- Page up
- Page down
- Home
- End
- Arrow keys
- Tab

A ... leafs through your manuscript one page down (further).
B ... moves the cursor to the end of a line.
C ... deletes characters to the right of the cursor as long as you press it.
D ... leafs through your manuscript one page up (back).
E ... moves the cursor to the beginning of a line.
F ... activates the overwriting mode, which means you can overwrite letters to the right.
G ... move your cursor in the directions given.
H ... stops a program or a command without losing data.
I ... deletes characters to the left of the cursor.
J ... produces capital letters as long as it is locked.
K ... both are used in combination with other keys for special functions.
L ... is the longest key on the keyboard, which produces a blank space when you press it.
M ... confirms a command; in a word processor it moves your cursor to the next line.
N ... moves the cursor to the right for a given number of spaces.
O ... produces capital letters as long as you press it.

b Draw up a list on a separate piece of paper and fill it in using the information from above.

Example:

number	name of key	description of function
1	ESC - escape	stops a program or a command without losing data.
2	Tab	

A2 The development of computers

a Complete the following text using the words in the box below. The text will tell you about the earliest developments in the history of computers.

mechanical calculating machines – devices – inventor – lessons – weighed – Englishman – century – modern – between – engineers

As with most modern devices or developments, the computer does not have one single ①. The names of people who paved the way for our modern communication ② date as far back as the 17th ③. In your science ④ you may have heard of Blaise Pascal or Gottfried Wilhelm Leibniz as the inventors of ⑤. Charles Babbage, an ⑥, invented a mechanical calculator in the 1830s.

The development of ⑦ computers in the 20th century is connected mainly with two names: The German Konrad Zuse built a programmable calculator in Berlin ⑧ 1936 and 1941. In 1944 the American Howard Aiken and a group of ⑨ working for a company called International Business Machines Corporation (IBM) developed an electromechanical machine known as the Mark I. Mark I was the first program-controlled calculator. The machine was more than 50 feet long and ⑩ 5 tons.

b Put the following paragraphs in the correct order and you will learn something about the next important events in computer history. Give a short summary of the text in French. The red letters in brackets form the name of an important milestone in the development of computers.

(C) In 1971 the first microprocessor appeared on the market, making the fourth-generation computers smaller and faster. This new technology made it possible to build small and lightweight microcomputers (PCs) that were affordable even for private homes.

(N) When vacuum tubes were replaced by transistors the development of computers got a new impetus.

(I) Transistors were invented in 1947 but were not used for computers until about one decade later. They made the units much smaller and also reduced the energy consumption. These second-generation computers were used mainly at universities and by governments.

(A) In the 1960s the invention of the integrated circuit revolutionised computer technology and started the era of the third-generation computers. Electronic elements such as transistors, capacitors, resistors and their interconnections were put together in integrated circuits (ICs) on small silicone wafers or chips, called microchips.

(E) In 1946 the American scientists John W. Mauchly and J. Presper Eckert completed their work on ENIAC, the first large-scale general-purpose electronic computer. ENIAC stands for Electronic Numerical Integrator And Computer. It was a first-generation computer based on vacuum tube technology and weighed 30 tons.

c Describe the performance of the first microcomputers and compare it with today's computers. What are the main differences?

A3 What type of computer do you need for your job?

Denise is a toolmaker. She has decided to go back to school to become an engineer. For the next two years she will be a student at Marvin Technical College. The teachers have told her that she needs a good computer to be able to do the required exercises. She has worked with computers before and she wants to update her own, but she is not quite familiar with the latest developments. At NuTech, a large computer retailer, she asks for assistance.

a Listen to the dialogue.
Take down the technical details on your notepad.

b Listen to the dialogue again, check your notes and answer the following questions on the text.

1. What does Denise want to do?
2. What kind of computer has she got?
3. What does the shop assistant think about Denise's computer?
4. What does Denise need the computer for?
5. Why does the shop assistant say that it is important to have a fast computer?
6. What does the shop assistant propose in the end?

A4 Comparing hardware

a Match the terms and features from the box below to the following hardware.

keyboard printer loudspeaker CPU monitor

ink jet – flicker-free – 50-sheet – paper tray – desktop – RAM speed – high resolution – print quality – ergonomically designed – 30 function keys – 30 Hz-20 KHz response

b Find out what kind of hardware is currently being developed? Use any sources available.

c **Compare the hardware and software of each of the advertised computers. Use the phrases in the box.**

Phrases: **Comparing computers**
• Model A has a built-in … • The main advantage of model B over model A is … • Both / All models are equipped with … • Model … comes with … • A new feature of model … is the built-in … • Model … is a low-priced alternative to … • Model … can handle almost any office task … • Model … is the tool for home offices. • Model … is designed to meet the needs of … (professional users / private users).

d **What is the target group for each advertisement? (private user / someone who likes to play games / a software programmer / a broker …)**

e **Find advertisements for computers on the Internet or in computer magazines and discuss them in your class.**

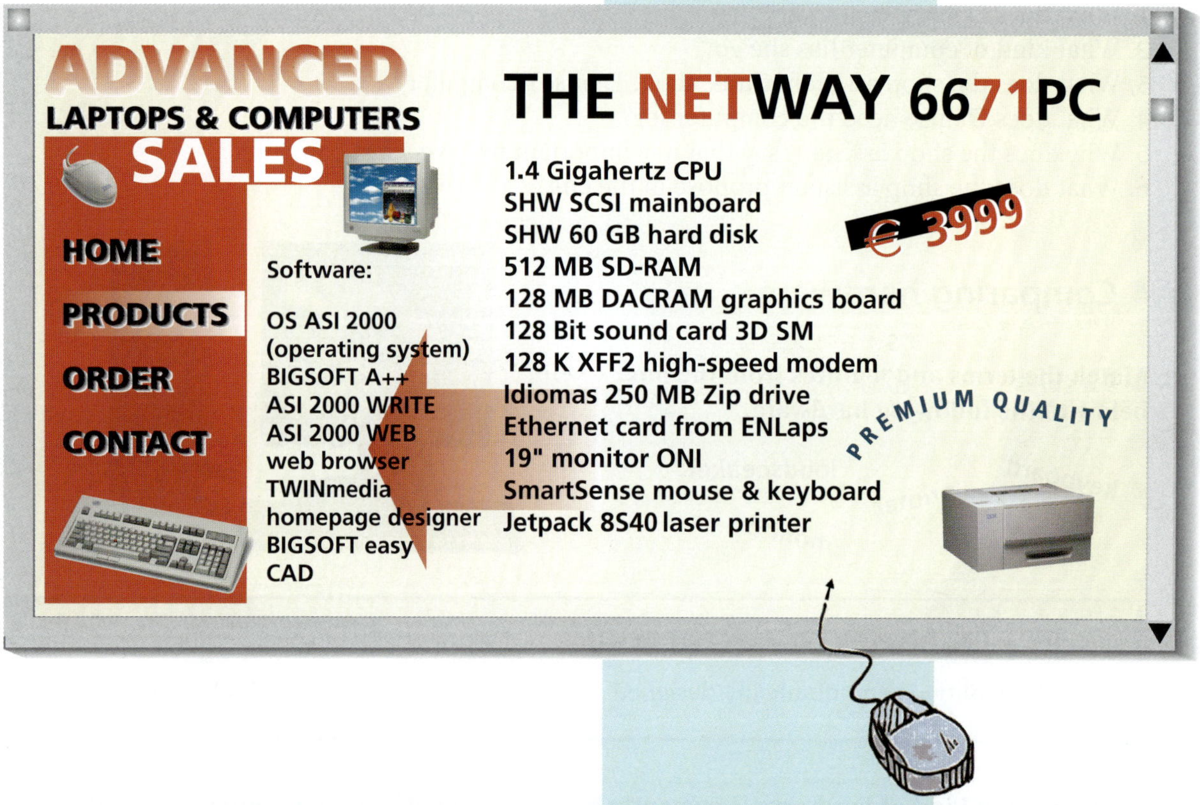

ADVANCED
LAPTOPS & COMPUTERS
SALES

- HOME
- PRODUCTS
- ORDER
- CONTACT

XENTUM Notebook

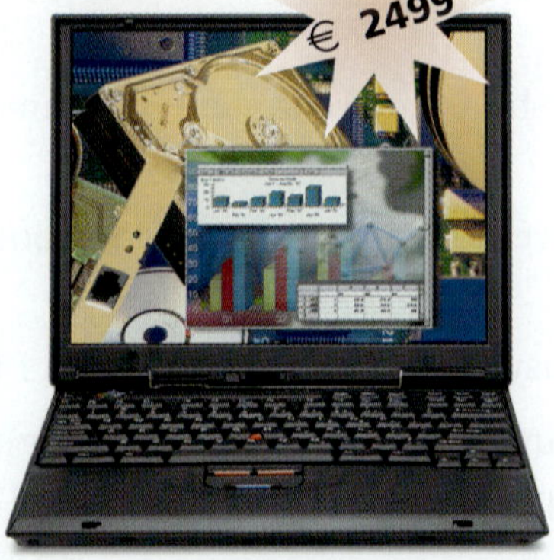

€ 2499

800 Megahertz CPU
20 GB hard disk
64 MB SD-RAM
32 MB graphics board
32 Bit sound card with speaker
XFF modem, featuring a socket to plug in your mobile if no normal phone connection is available
DVD-ROM drive
15,1" display

Software: OS ASI 2000 operating system, ASI 2000 WRITE, ASI 2000 WEB, web browser, MODura e-mail software

ADVANCED
LAPTOPS & COMPUTERS
SALES

- HOME
- PRODUCTS
- ORDER
- CONTACT

The GetReady PC

Software:

OS ASI 2000 operating system
ASI 2000 WRITE
ASI 2000 WEB
web browser

1 Gigahertz CPU
SH mainboard
SH 22 GB hard disk
128 MB SD-RAM
32 MB DACRAM graphics board
128 Bit sound card 3D with speakers
XF high-speed modem
16x DVD drive SH
17" monitor ONI
SmartSense mouse & keyboard
Jetpack 8S31 colour printer

€ 1199

SPECIALS

SEARCH

A5 Troubleshooting: Problems with installing hardware

At the help desk of Eastshore Entertainment, a software company that designs computer games.

a Read the following dialogue in pairs and describe how the help desk is treating the customer.

Help desk	Hello, this is help desk 17 at East-shore Entertainment; my name is Linda Euston. Can I help you?
Customer	I've been trying to reach you for over ten minutes! How long does it take to pick up a phone?
Help desk	I'm really sorry, but we are very busy at the moment. How can I help you, sir?
Customer	I've got this game called "Invade and Take Over". I'm having problems with the 3D sound. The game doesn't accept the default settings of my 3D sound card.
Help desk	So, you have no sound at all?
Customer	All I can hear is a screeching sound.
Help desk	I see. I can understand that this is very annoying. Which version of "Invade and Take over" do you have, sir?
Customer	I think it is the 3.0, second release. Do you need my serial number?
Help desk	No, thank you, that won't be necessary. What I do need to know is what kind of sound card you have.
Customer	I have a "3D sound card" from a company called "Creative Tech".
Help desk	Oh yes, we've had several problems with this sound card since the release. The problem's quite easy to fix. Do you happen to have access to the Internet?
Customer	Yes, I do.
Help desk	This is what you should do: Go to our Website; the address is www.Eastshore.com (the address is also on the last page of your manual). Click under "downloads" and select "patchCT" for Creative Tech.
Customer	Does it take very long to download?
Help desk	That depends on your connection, but it shouldn't be any longer than 10 minutes.
Customer	And you can guarantee that it works?
Help desk	We've never had any problems so far. If it doesn't work, you can call us at any time on 1 800 555 267. Anyway, please give me your address; we would like to send you the "Invade and Take over" mousepad and T-shirt free of charge to make up for the problems you've experienced.
Customer	Oh, that's very kind of you, and thank you for your help.
Help desk	You're welcome. It was a pleasure to serve you, and please don't hesitate to call again if there should be any other problems.
Customer	Thank you, I will.

b Describe in your own words: What is the customer's problem?

c Role play: Student A is the calling customer, Student B is the help desk representative.

Student A (customer):
Angry, because he / she has had to wait so long for the phone to be answered / problem with the new computer he/she has bought / computer doesn't boot / it says on the screen, "no system disk, boot error" / says this problem was caused by the company.

Student B (help desk):
Tries to calm customer down / asks when computer was bought / asks what he/she can see on the screen after switching it on / asks him to check whether there is a 3.5-inch disk in the computer / asks him/her to take it out / asks if it works / apologises for not having taken out the disk when preparing the computer in the shop before it was sold / makes an offer to the customer to call into the shop and pick up a CD-ROM with fonts for free.

d Find out what is wrong. Match the problems (1.–10.) with their possible solutions in the box below.

1. Paper jam.
2. Printer doesn't work.
3. E-mail is returned.
4. Program cannot save file on disk.
5. When you start your computer it says "No system disk" and stops.
6. Your word processor cannot open a file.
7. You cannot get access to the Internet.
8. Your system program tells you that it cannot start a program.
9. You play a multimedia game and cannot hear the sound.
10. Your computer cannot read from your CD-ROM drive.

Possible solutions:
Check if your modem is switched on and properly connected. • Not enough memory, too many programs are already open. • Remove floppy disk and hit any key. • Either disk is full or the write protection is enabled. • Check the plugs. • Open the lid of the tray and remove the jammed paper. • Check for correct address: every dot is important. • CD-ROM needs cleaning. • File is already opened or corrupted. • Reinstall your sound card.

A6 Storage devices: Floppy disk – Hard disk – CD-Rom – DVD

Files can be stored either on your hard drive or on a hard disk. Hard disks consist of a stack of metal disks in a separate frame inside the computer. Their storage capacity is many times higher than that of a floppy disk. Hard drives store and retrieve data from spinning hard disks.

They do not last forever and many people have witnessed crashes of hard disks and data being lost. It is essential, therefore, that you save your files on other media as well. One of the most common mediums used to be a floppy disk. It has a storage capacity of about 1.4 MB, which is not much, but enough for storing files of your word processor. Technical drawings, however, require more storage space. Here, a zip-drive with 100 MB or 200 MB storage capacity would be more suitable. A zip drive can also be used to make backups of your files.
Even more storage space is provided by a CD-ROM, which is the storage medium used for most software that is available on the market. It has a storage capacity of 650 or 750 MB. The most promising storage device for the future, however, is the digital versatile disk, DVD in short. It has an enormous storage capacity of up to 17 GB, which is enough to store entire movies.

a Translate the text into French.

b Using the information from the text above draw up a list on a separate sheet of paper and compare the different storage devices. Use information from the Internet, computer magazines, etc. to update the list.

Storage medium	storage space	price (if available)	use
hard drive			

B Software

System software translates the commands of the computer into a language that the peripheral devices understand. It usually runs automatically without the user even noticing it. Application software, on the other hand, is designed to perform certain tasks such as writing, calculating, painting or drawing. Application programs cannot be started without the help of system software.

a Give a definition of "software" by putting the two sentences in the correct order:

1. tells / the hardware / what to do / is / a set of instructions / software / that /.
2. there are / system software / two kinds / software: / of / applications software / and.

b Decide which of the following statements are true for system software (s) and which for applications software (a):

1. controls the computer's internal operations
2. translates the user's commands into computer language
3. includes word processors, spreadsheet programs or database management
4. controls peripherals such as monitors or floppy disk drives

c Give examples of both system software and applications software. Use current advertisements, catalogues, the Internet or any other source.

B1 Drawing pictures – CAD

There are various paint programs: The simple ones allow you to draw colourful pictures that you can print out or integrate into a piece of text. Or, if you are skilled enough, you can create illustrations for your presentations. The more refined ones are for technical drawings that enable you to design complicated parts on the screen.

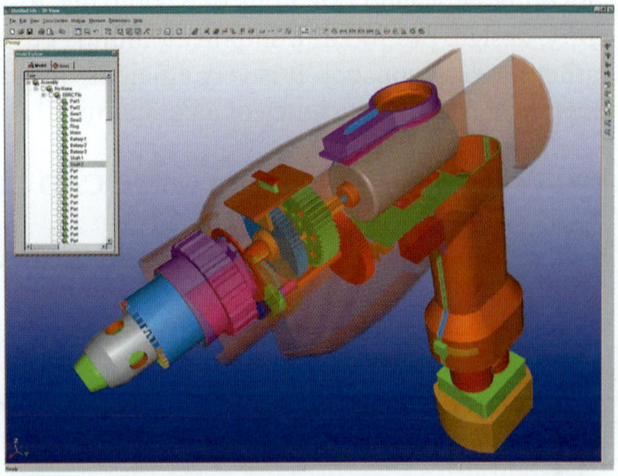

- **Match the words from the box with the gaps (numbers match letters) on a piece of paper:**

(a) complicated (b) measurements (c) parts (d) versa **(e) aided** (f) angle (g) people (h) printed (i) models (k) attachments (l) possible (m) technical

Example: 1e (computer – **aided**)

In business, especially in the field of technical drawing, you have to use more refined and powerful programs such as those for CAD-programs for computer-① design. These programs allow you to create ② drawings of ③ machine parts, for example, on any given scale. The images created are 3D ④, and this is an enormous advantage; these programs make it ⑤ for the engineer or architect to look at his design from almost any ⑥.
New machine ⑦ can thus be designed on a screen, and the finished drawing can be ⑧ out with the exact ⑨ and explanations. Another advantage of electronically designed plans is that they can be sent as e-mail ⑩ via the Internet from one continent to another, from the design office of a company in France to its manufacturing unit in England or vice ⑪.
⑫ from all over the world can work on the same project 24 hours a day.

B2 Computers in industry – CAM and CIM

Apart from CAD, there are many more uses for computers in industry. Computers run robots or control automated assembly lines, they are excellent planners and organisers of complicated production processes, and they keep track of materials in stock. This wide field of computer applications is called CAM – **C**omputer-**A**ided **M**anufacturing.
CIM (Computer-Integrated Manufacturing) goes one important step beyond CAD and CAM, which centre on single and isolated steps of

production, such as the technical design of a product or its automated machining. CIM integrates all components and functions of a company, from the first contacts with customers, to offers, order entry, design, production, shipment, billing, customer service and even after-sales service. CIM is a rather modern concept. It is very efficient, because it keeps control of every aspect of business and thus increases the efficiency of a company.

a Give definitions in your own words: CAD – CAM – CIM.

b Give examples of how computers are used in industry.

B3 Using a computer to write: A word processor – important features

Using a **word processing** program to write makes it easy to experiment with your ideas. Editing documents is very simple using the cut-and-paste features. With these, you can easily move words or blocks of text around and experiment with the changes. This is quite helpful, for instance, in translating brochures or manuals.

Besides helping you to follow your creative impulses, word processing programs do some tedious tasks, such as proofreading, almost instantly. They let you make changes without having to retype the entire page. A **spell checker** uses a dictionary of stored words to check the words you've written. If it locates a misspelling, you can ask for a suggestion to correct it. You can use a grammar checker to help locate errors. Correcting grammar is more difficult for a computer than correcting spelling, however, because there are a lot of rules and many exceptions to them.

a Sum up the main features of a word processing program.

b Match the English verbs below with the French translations on the right.

copy – cut – edit – format – insert – paste – save as – search – number – retrieve – select – undo – replace

A. couper
B. remplacer
C. copier
D. annuler
E. formater
F. coller
G. rechercher
H. enregistrer sous
I. numéroter
J. éditer
K. insérer
L. sélectionner
M. récupérer

c Match the French nouns below with the corresponding English translations on the right.

caractère – fichier temporaire – fichier – coupure de mot – icône – commande – fenêtre – champ de dialogue – taille de police – titre de bas de page – en-tête – vérificateur d'orthographe – aperçu

A. file	H. dialog box
B. character	I. font size
C. temporary file	J. footer
D. icon	K. header
E. view	L. hyphenation
F. window	M. spell checker
G. command	

d Modern word processors have become faster and faster. They are so complex that they can do almost anything. Nevertheless, the normal user – that covers an estimated 80% of all possible users – takes advantage of only 20% of the features provided. **Read the following sentences and find out about these features by combining the beginnings of the sentences on the left with the corresponding endings on the right. Write the complete sentences on a separate piece of paper.**

a Basically, all word processors

b Most of them follow the WYSIWYG standard;

c For example, if you format a headline in bold face or italics

d One of the greatest advantages of a word processor is

e The cut-and-paste feature is very useful

f Another useful feature is the drag-and-drop function that makes it possible for you

g You can also put pictures or drawings into your text

h A helpful feature is the spell checker

i If you cannot think of the right word

j A lot of word processors even allow you to create your own WebPages

1 that you can edit the text (change or correct your text) as often as you want.

2 as you simply cut out a word where you don't want it and put it back in some other place where it fits better.

3 work the same way.

4 that automatically checks your text for spelling mistakes.

5 you see it on the screen almost exactly the way it is printed out later.

6 that means, what you see on the screen is what you get as a printout.

7 to pick up a word or an object with your mouse, pull it across the text and simply drop it wherever you want it.

8 and you can fit your text around these pictures or drawings.

9 that you can upload directly onto your website.

10 the thesaurus may be quite useful to you.

e Complete the following text using words from the previous pages.

Like paint programs, a word ① is an application program. Among its leading features are the cut-and-② or ③-and-drop functions. An advantage over writing with a ④ on a piece of paper is that you can ⑤ your text on the ⑥ and ⑦ it out when every mistake has been corrected. You can even add in ⑧ or ⑨ and fit the ⑩ around them. A very helpful feature is the ⑪ that can automatically point out spelling mistakes and even ⑫ them automatically if you wish. But you should not trust it too much; your knowledge of orthography is still needed. Also very useful is the ⑬, that can suggest alternative words. For example, if you have used the word "ask" once or twice before and want to avoid it a third time it may suggest using "to question, to request, to inquire, to interrogate, to query" instead. What is also nice is that you can give your text a personal look by choosing special ⑭ such as Times Roman, Century Gothic, or Old Western. Modern word processors provide another ⑮ for Internet users: you can transform your text into html-text and upload right away onto your ⑯.

B4 Spreadsheet programs

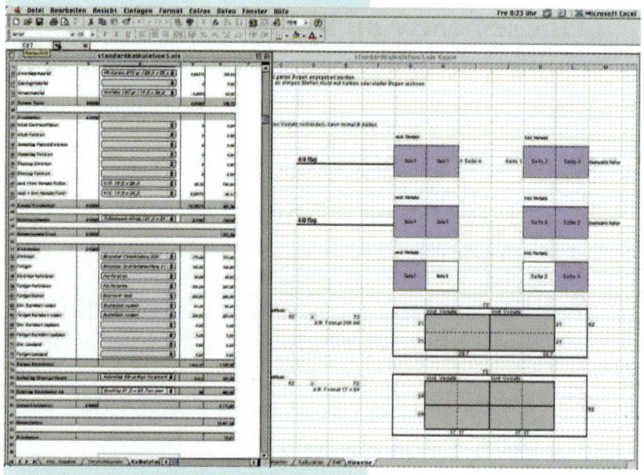

Spreadsheet programs are another type of applications software. They are similar to word processors – you can even use a spreadsheet program to write texts – but their main purpose is to calculate and present numbers and data in a table format. When you open a spreadsheet program you will notice that the screen is divided into rows and columns. In each of the boxes created by these horizontal and vertical lines you can put in either letters, words, sentences or numbers, even complete mathematical formulas. Thus the user can tell the program to do all possible mathematical calculations with the numbers, in the boxes above or to the left or anywhere within one or more spreadsheets. For example, you can add up figures, divide them, or work out a certain percentage. You can also transform the numbers into graphs that can be presented to the reader, perhaps to illustrate the development of a company's turnover. As with a word processor, you can format your printout, e.g. by choosing different fonts.

- **Answer the following questions and write down the numbers that correspond with the answers you have chosen. If you put the correct numbers together you'll discover when the first spreadsheet program was developed.**

1. Can you use a spreadsheet program to write a letter? Write down 1 for "yes", 0 for "no".
2. Can you only do mathematical operations with numbers on just one sheet?
   Write down 8 for "yes", 9 for "no".
3. Can you use only one font? Write down 6 for "yes", 7 for "no".
4. Can you insert a line graph or a bar graph into your spreadsheet?
   Write down 6 for "yes", 7 for "no".

B5 Organizer manual

a Read the following installation manual of a desktop organiser and complete the text with the words in the box below.

automatically – corrupted – saying – loaded – entry – message – double-clicking – view – manually – deleted – copy – options – receive – entered – save – button – search

Installation (A1)

Start the installation program by ❶ on the file setup.exe. Follow the instructions on the screen. Then restart your computer ❷. If everything goes right, the Organizer window should pop up every time you start your computer.

Deinstallation

Simply start the setup program again and click on "uninstall". Note that all of the files are ❸. If you want to keep your entries, you will have to ❹ them from "C:\orgnz\data" to another directory.

The main window

From here you have the following ❺ (click on a word to ❻ an explanation):

Create a new entry (A2)

If you click on this ❼ a new window opens. First enter the date and the month concerning the entry you wish to make. Then click on "Enable writing". This is done to ensure that an entry concerning that date doesn't yet exist. If it does, you will receive a message saying so; you can either change it or view it. If the file doesn't exist yet, writing will be enabled. After you have ❽ your text, click on "❾ entry".

Change an entry

First enter the date and the month for the ❿ you wish to change. Then click on "Load entry". If the entry you wish to change doesn't exist you will receive a ⓫ saying so. If it does exist, the entry is loaded and writing is enabled. Now you can make your changes. When finished, click on "Save entry".

View an entry

First enter the date and the month for the entry you wish to view. Then click on "Start ⓬". If the entry you wish to view doesn't exist you will receive a message saying so. If it does exist the entry is loaded and you can ⓭ it.

View today's appointments

The entry is loaded ⓮. If it doesn't exist or if it is ⓯ you will receive a message saying so.

View tomorrow's appointments

The entry is ⓰ automatically. If it doesn't exist or if it is corrupted you will receive a message ⓱ so.

Delete an entry (A3)

Enter the date of the entry you wish to delete. Then click on "Delete entry". Note that once an entry is deleted it can't be restored.

(A1)

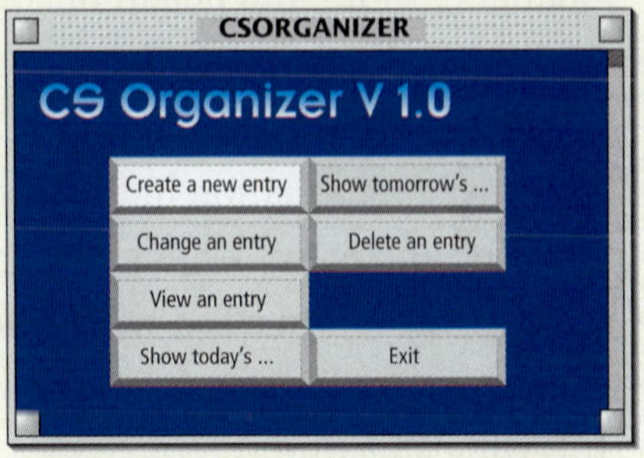

(A2)

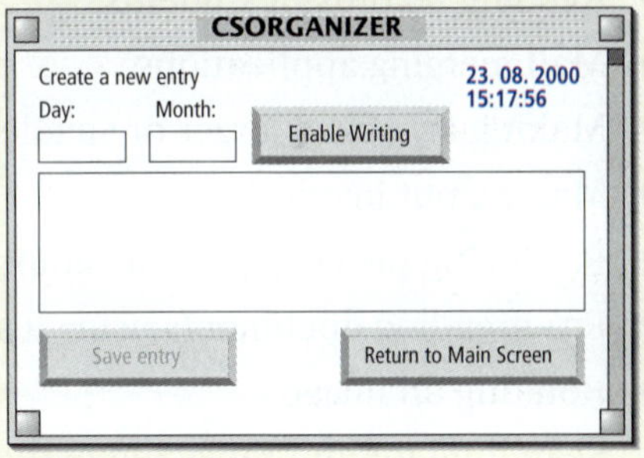

(A3)

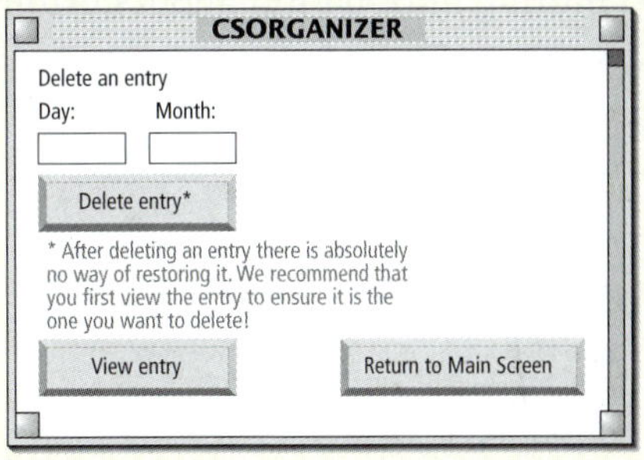

b **Work in pairs. Student A: Ask questions about the organizer manual on page 47. Student B: Answer the questions.**

Example:

Student A:
How do I install the program?

Student B:
First, you click on setup.exe.
Then you ...

B6 Programs and applications

- **Which of the programs on the right can do the jobs below?**

- Adding colour and patterns to objects
- Automatic numbering of chapters and sections
- Calculating interest rates
- Changing the value of cells
- Changing width and height of columns and rows
- Controlling all sorts of data
- Creating rectangles, squares or other shapes
- Displaying information in the form of a table
- Drawing and manipulating objects
- Exchanging information with other computer users
- Formatting multiple columns of text
- Inverting a picture
- Keeping records of students
- Mail merging applications
- Making an object larger or smaller
- Making out invoices
- Producing pie charts or line graphs
- Proofreading documents using a spell checker
- Rotating an image
- Searching for and replacing particular words
- Storing information about customers
- Writing faxes
- Writing reports

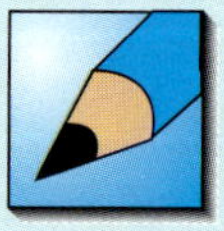
graphics program(6x)

word processor (7x)

spreadsheet program(6x)

database (3x)

electronic mail

C The Internet – Communication and information

a How do people communicate with each other?
Name the different means of communication that are in use today.

b Put the jumbled sentences in the correct order to discover what communications devices some people use and why.

1. business people / mobile phones / on trips / often / use / so / on public telephones / rely / they / don't have to
2. with / popular / are / more and / young people / mobile phones / becoming/ more
3. go /overseas calls / usually / satellites / via / because they / than undersea cables / faster / are
4. for / ordering / spare parts / use / engineers / often / because / a fax machine or the e-mail / that way / can / send / they / of the part they need / an exact drawing
5. postcards / in use / for sending / Christmas greetings / traditional / are still

C1 Getting information

Our age is often called the age of information. Is this because everybody and every institution gathers as much information as possible? Or because we can get information on almost anything under the sun? Or because we have so many sources for the right information? Don't you sometimes feel that we are drowning in information we do not need?

a Discuss the picture. Use the phrases below.

The picture shows / the picture expresses / the picture displays …
I feel exactly / don't feel this way …
The picture may exaggerate a bit …

b Look at the following list of information sources and your possible points of interest. Explain which source you would use and why.

To find out about … I would use …

The best choice here is definitely … because it is fast / cheap

In this case I personally use / I personally prefer to use … because it is reliable

I would not use …

It is surely a good idea to use …

Sources of information
telephone
newspapers
television
radio
weekly magazine
special interest magazine
information brochures from companies
encyclopedia in 20 volumes
encyclopedia on CD-ROM
dictionary in 2 volumes
dictionary on CD-ROM
specialised book / technical book
Internet

Points of interest
the latest football results
the meaning of an English word
the invention of the steam engine
the DAX
the production of steel
the date of your next classroom test, which you have forgotten
a problem you have with your printer
tomorrow's weather
the melting point of copper
a consumer report about a new camera
a new method of hardening metals

c Choose two or three of the sources of information mentioned on page 50 and tell your class about their advantages and disadvantages. Think of costs, availability, reliability of the information, accessibility of the source and whether is up to date.

Examples:

A has many advantages over **B**. It is cheaper and more reliable.
A disadvantage of **B** is that it cannot be updated ...
A definite plus of **A** is that ...

C2 Databases – Electronic sources of information

Not long ago the average family often had 15 or 20 volumes of an encyclopedia on their bookshelves in the living room, and these nicely printed and heavy books sometimes served as the family's source of information for a lifetime. Today, as our knowledge of the world doubles in only a few years, we are relying more and more on databases which are constantly updated, thus providing information even on the latest technological developments. In contrast to the databases, a traditional encyclopedia is already outdated on the day it is published.
So the most up-to-date sources for our knowledge – be it in technology, medicine, physics or even history – seem to be electronic databases. They can be published on CD-ROMs, on DVDs or in the Internet. Not only can they be updated in a very short time, but they are also often much cheaper than books. There is no need to read whole entries for a single piece of information. That is done by a search engine. In the end, you download only the information you actually need. You can then print out your material and take it with you or incorporate it into your text. Or you can mail it to somebody who needs it. Modern encyclopedias on CD-ROM make use of multimedia technology which allows you not only to read about a topic, but to hear about it or even watch short animated sequences. If you are interested in certain musical instruments, for example, you can hear what they sound like while reading about them.

a Using the information from the text above discuss the advantages / disadvantages of electronic databases over traditional books or encyclopedias.

**b Find information about the terms below in either traditional encyclopedias or on CD-ROMs or the Internet and present your findings to your class.
Which sources were the most effective?**

- thermosetting plastics
- greenhouse effect
- iron
- alloy
- electricity

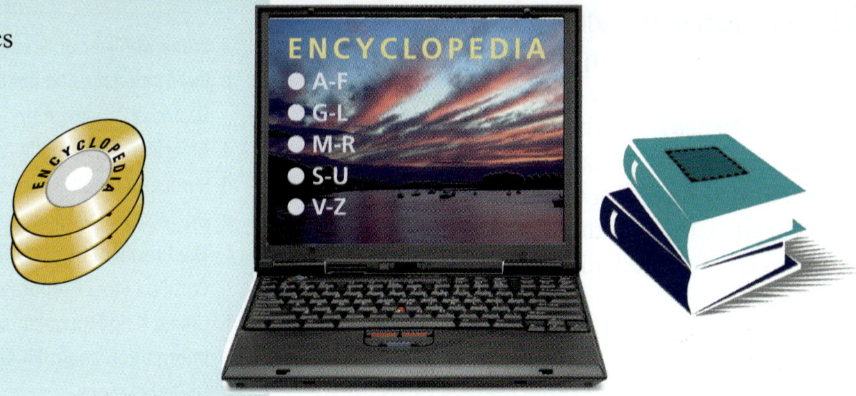

C3 The Internet

What originally began as a network for the US defence system has turned into an immense global network. During the Cold War, military experts were afraid that their information system could be too easily interrupted by destroying just one station. So they developed a network in which a single station could be reached through various paths. That was in 1969. Later, governments, administrative offices, universities and research institutes were also allowed to use the net.

At the beginning of the 1980s, the Internet as we know it today already connected thousands of private and public networks all over the world. But yet another invention was necessary without which the Internet would never have become such a success: the most important service within the net – the World Wide Web (www) – with a common protocol to enable communication between servers and clients. The so-called hypertext terminal protocol (http) was invented in 1989 by Tim Berners-Lee together with other scientists and CERN (European Organisation for Nuclear Research) in Geneva, Switzerland. Thanks to this protocol, we have a common standard with which we can publish and search for text or pictures in the web, as it is commonly referred to.

a Summarise the brief history of the Internet. If you can, give more facts from your own knowledge, from encyclopedias or other sources.

b If you have had any experience with the Internet, tell the class about it. If you haven't had any, tell the class why not (no modem, no computer that is fast enough, or simply because you are against the Internet). Give reasons.

c What are the latest developments regarding the Internet?

C4 Internet addresses

With the information from the text above it is quite easy to explain the structure of a typical Internet address, for example http://www.ft.com where *http* stands for hypertext terminal protocol and *www* for World Wide Web. *ft.com* is the name of the domain (for the *Financial Times,* in our example), *ft* being the individual part of the address and *com* the ending of addresses for large companies. There are other endings as well, such as *org* for organisations, *edu* for colleges and universities in the USA, *gov* for websites of governments and their institutions in Great Britain or the United States, *fr* for addresses in France, *de* for Germany, *uk* for the United Kingdom.

- **Which Internet addresses (1–10) will offer you information about the subjects (a-k)?**

1 http://www.whitehouse.gov
2 http://www.carolinasteel.com
3 http://www.number-10.gov.uk.
4 http://www.open.gov.uk
5 http://www.city.philadelphia.com
6 http://www.un.org
7 http://aaa.com
8 http://www.virginia.edu
9 http://www.iccwb.org
10 http://europa.eu.int/comm/dg05/elm/eures/index.htm

a Prime Minister of Great Britain
b International Chamber of Commerce – World Business Organization
c American Automobile Association
d European Employment Service
e University of Virginia
f The President of the United States of America
g Carolina Steel Corporation
h United Kingdom public sector information on the Internet
i Information about Philadelphia
k United Nations

C5 E-mail

Another very important use of the Internet is e-mail. Hardly any business or company can do without it. E-mails reach their destination within seconds or minutes, no matter where you send them. The e-mail address also follows a certain structure. What is new in comparison with a website address is the sign @ (meaning "at", as in the preposition "at"). chris.myers@fantasy.net gives you to the left of the @ sign the e-mail name of a Chris Myers, and to the right of it the network or provider where Chris Myers has his e-mail account.

- **Ask your classmates about their Internet addresses. Find out about the different providers used and, if possible, why people choose a special name or alias for themselves.**

C6 The Internet – some legal aspects

Everything seems to be available on the Internet; it is so easy to download almost every picture you see, every text you would like to have – and yet, unless otherwise stated, there is a copyright on every piece of text, every graph, every piece of information. Many kids have felt the long arm of the law when they put pictures of their film heroes on their private homepages, have even heard from lawyers of their favourite soccer clubs after they reprinted fan items on their websites. Everything published in the web is protected by international copyright laws, just as is all information in magazines or books.
There is another warning you should always keep in mind when you are on the Internet – don't ever be too sure about the safety of your own data. While you are on the net somebody else might check out your hard drive, find out your passwords or even see whether you run unlicensed copies of programs. Be very careful when you buy something on the net and you are asked for your credit card number. There is no absolute safety from being spied upon by sly people.

a What dangers do you risk if you use the Internet, according to the text?
What dangers does the text not mention?

b What precautions can you take to avoid problems while being online?

- always use a virus protector
- never give your credit card number to companies you do not know
- if possible, use a separate computer for the Internet
- do not keep personal data on computer
- never surf after you have checked your bank account online
- never leave computer alone while online

C7 Information management

The Internet is a bottomless pit, someone has said, when it comes to information. Users often complain that they receive and download too much information, that they do not have the time to read everything they've got. What is necessary is information management.
This involves two things:
First, download or keep only what you are absolutely interested in. Otherwise you will drown in information. Second, be very specific about what words you put into search engines when you are looking for information, in order to exclude information that is not needed.

a What can you personally do to avoid being drowned in information?

b Experiment with search engines on the Internet or with encyclopaedias. Read their instructions carefully.

Here are a few hints that might help. For example, if you want to find something about the history of computers, don't just type in "computers", give the word "history" as well. The same holds if you are interested in "foamed metals"; do not search just for "metals", but give a second keyword in addition.

c If you have access to the Internet, and if you know how to use search machines, you can look for information in English about companies in your area. Present the information gathered to your class.

C8 The Internet language

- **Find words and abbreviatons related to the Internet in the word square.**

B	T	O	O	L	B	A	R	K	H
C	R	H	W	A	G	J	M	O	Y
M	D	O	W	N	L	O	A	D	P
O	O	M	W	E	B	S	I	T	E
D	M	E	H	S	U	R	F	E	R
E	A	P	T	I	E	K	S	V	T
M	I	A	M	S	Y	R	W	A	E
J	N	G	L	D	L	I	N	K	X
V	Q	E	O	N	L	I	N	E	T
C	H	A	T	B	O	X	U	S	B

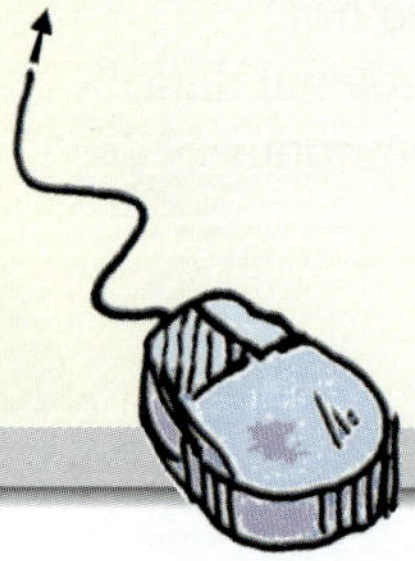

C9 Multimedia

Multimedia is believed to be the future of all computer applications, a most astounding combination of text, sound and motion pictures, together with interactive components that can be used to influence the sequence of what is presented on the screen and the loudspeakers. The forerunners were the computer games that produced sequences of film with real sound in front of the player's eye and gave him the chance to decide which steps the heroines or heroes of his games should take.
More serious applications nowadays are language programs that – for example – present a certain grammatical problem and give in boldface letters the rules to follow; in addition, you can hear the sentences spoken by a native speaker. You have to fill out forms with exercises to show that you understand the rules. By reading your sentences into a microphone, you can also have your pronunciation automatically checked and corrected by the program.
Companies use multimedia to present their enterprises to customers and shareholders: a beamer connected to the computer projects on the wall the headlines of a presentation accompanied by the sound of speech, music or factory noises. Even film sequences can be included. The impression produced by this new technique is very powerful because it uses many different media to conquer the senses of the audience.

a **Define multimedia using information from the text above.**

b **Which applications of multimedia does the text describe?**

c **Describe any experience you may have had with multimedia.**

d **What do you think will be future uses of the computer (e.g. computers controlling your electricity system at home, computers keeping track of your shopping list or your appointments, computers guiding your car on a business trip, handling hotel reservations, etc.)?**

D Grammar revision: Auxiliaries

See Grammar Overview in the appendix.

a Put in suitable auxiliaries: can/cannot, must/must not, need/need not, should/should not.

Example:
You **can** use your rewritable CD-ROMs and DVDs repeatedly if you treat them right, but you **should** always bear in mind that they are not indestructible.

How to treat your CD-Roms and DVDs

- *You ① use your rewritable CD-ROMs and DVDs repeatedly if you treat them right, but you ② always bear in mind that they are not indestructible.*
- First of all, you ③ expose them to extreme heat or cold. Store disks in a cool and dry place.
- You ④ keep them in an air-conditioned room, but the room temperature ⑤ be too high.
- You ⑥ store them in a protective case to avoid scratches on the surface of the disks.
- You ⑦ use abrasive or solvent cleaners as these can damage the disks!
- You ⑧ touch the surface of the disks!
- DVDs ⑨ be handled even more carefully than CDs as they can be easily damaged by bending or twisting.
- Disks ⑩ be cleaned with a damp, non abrasive, lint-free cloth.
- Remember that disks ⑪ be damaged if they are stacked. So never stack them.
- If you don't follow the above rules your disks ⑫ work properly.

b Rewrite the following sentences with the alternatives given in the box on the right. Note that a different tense is necessary, but the auxiliaries cannot form the new tenses themselves. Use the alternatives instead.

Verb	Alternative
can =	to be able to
must =	to have to
needn't / need not =	don't have to / doesn't have to
may =	to be allowed to

Example:
He **must** contact his software dealer. Yesterday he ? contact his software dealer.
Yesterday he **had to** contact his software dealer.

1. He must contact his software dealer. If the program doesn't run tomorrow he ? contact his software dealer.
2. The people there can certainly help you. Often in the past they ?
3. Fred: "Imagine, in the camera shop they said: *You may take the digital camera home with you over the weekend and try it out.*" – Cindy: "So you ? to test the equipment last weekend?"
4. You don't need to install the software again. The last time this happened we ?
5. Part of the assembly line must be reorganised because they are changing the software. They found a mistake in their software, so part of the assembly line ? be reorganised last week.

Unit 4

Mechanical engineering – Tools

There are two main groups of tools: hand tools such as a hammer, and power tools which are driven by electric power or compressed air. Power tools are also split into two groups: manual power tools such as an electric drill, and machine tools such as a column drill.

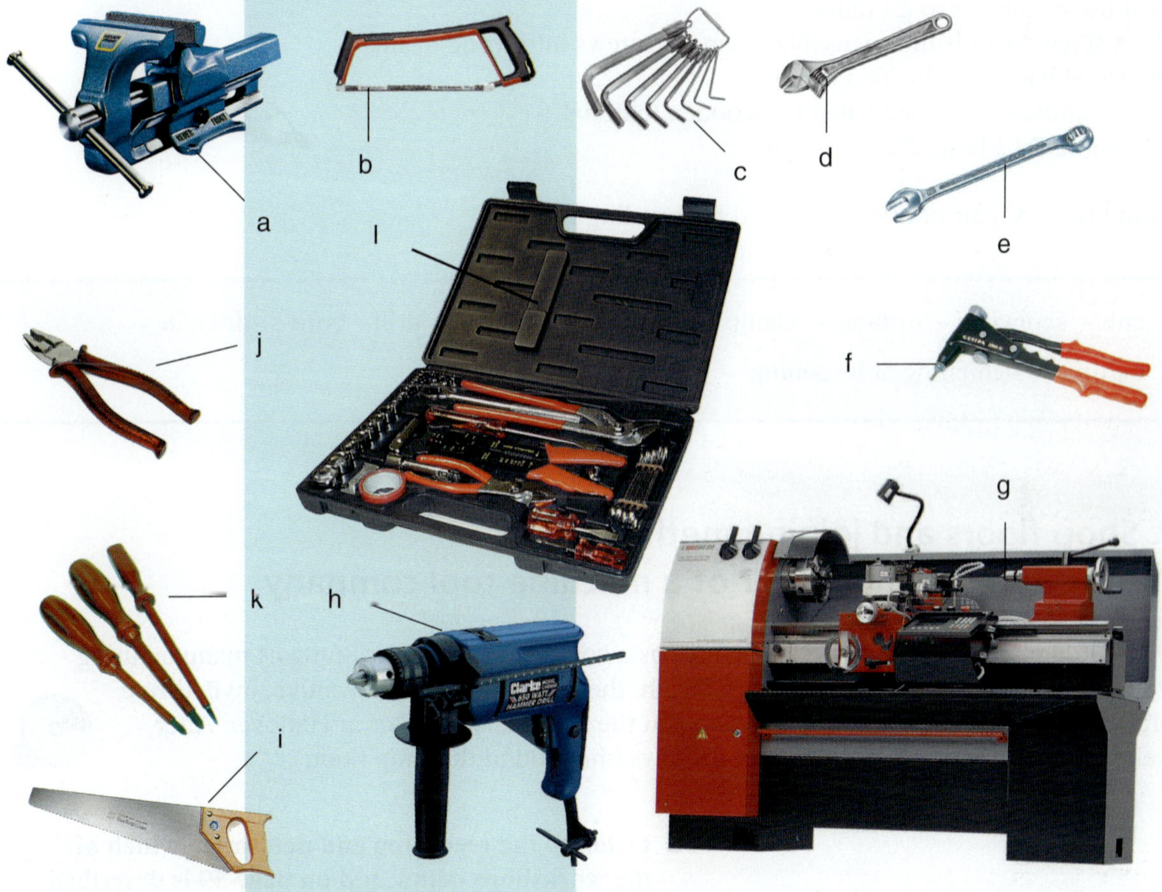

a Match the following tools with the illustrations above.

1. screwdrivers
2. combination pliers
3. spanner
4. set of Allen keys
5. lathe
6. vice
7. hacksaw
8. handsaw
9. electric drill
10. adjustable wrench
11. riveting tongs
12. toolbox

b List the above tools under the following groups:

- **machine tools**
- **power tools**
- **hand tools**

c Complete the following sentences using words from the box below

What is a ... used / needed for?

1. I use a spanner for ____ bolts and nuts.
2. With a column drill you can ____ holes up to a diameter of 10 mm.
3. A centre lathe is used to ____ round/cylindrical materials such as bars or shafts.
4. You need a bench vice to ____ a workpiece for filing, deburring or bending.
5. Horizontal and vertical milling machines are used to ____ flat materials with different forms.
6. I use a hammer to ____ nails.
7. A screwdriver is indispensable when ____ screws into wood or tin.
8. Metal bars are ____ with a hacksaw.
9. A handsaw is needed when ____ wooden material.
10. Before ____ a hole, ____ its location.

Example: A pair of pliers is needed for bending wire.

cut – screwing – drilling – clamp – mark – work (2x) – drill / bore – drive in – cutting – tightening or loosening – ***bending***

A Shop floors and joining methods

A1 Visiting the shop floor of a machine tool company

DEP, Design and Engineering Performance, is a company based near Stuttgart manufacturing state-of-the-art machine tools. Christof Braun, the company's chief executive, invited Bob Morris to visit the company when they met at the last Industries Fair in Hanover. After welcoming his visitor to DEP Mr Braun shows him around the shop floor.

a Listen to the recording and determine which of the workshops below and on page 59 is described in the dialogue. Give reasons for your answer.

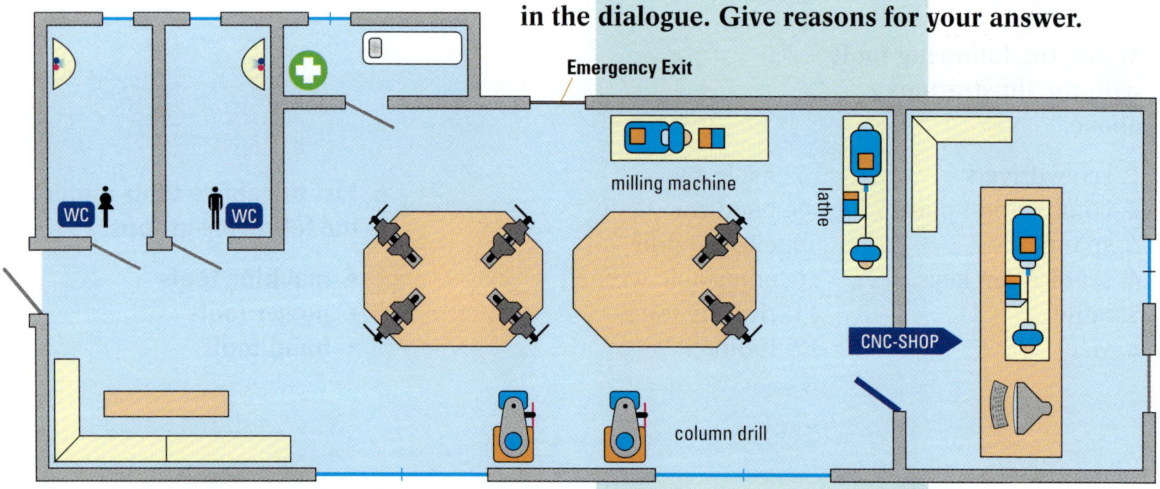

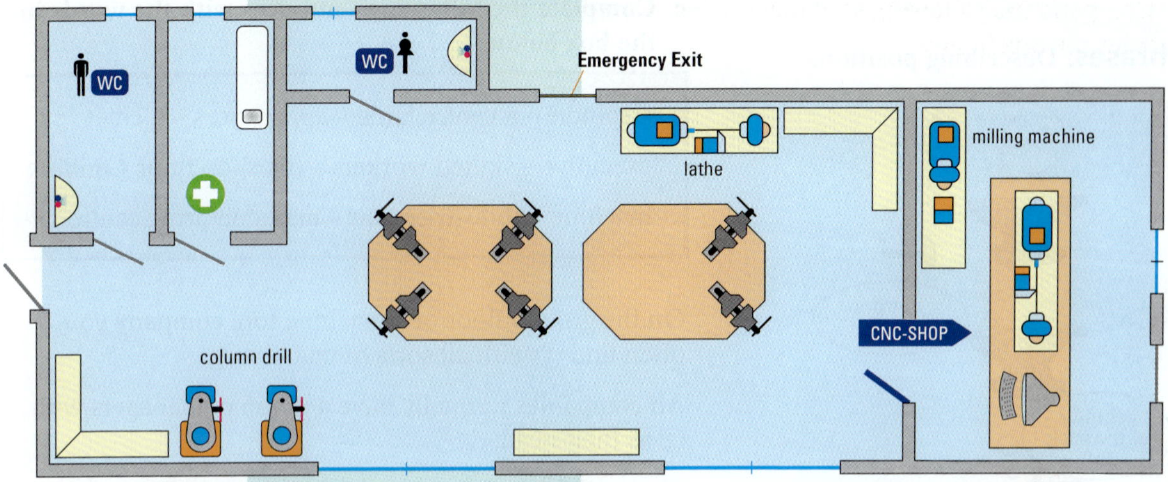

Christof: "This is our general workshop, where our apprentices are trained in mechanical engineering. In the centre are the work benches with a vice on each bench plate."

Bob: "And between the large windows over there I can see two column drills, and there is also a centre lathe next to that blue door, isn't there, and a milling machine opposite the lathe?"

Christof: "Yes, that's right. Some of our training units include basic machine training, too, so our young apprentices get their first experience with the machinery they'll be operating later as skilled workers."

Bob: "I see. Now, could I have a look into your CNC shop, Christof? It's the heart of your manufacturing centre, isn't it?"

Christof: "Yes, you're right. It's just behind that blue door. That's where you'll see our latest lathe model with a high-speed spindle doing up to 10,000 rpm. Do you remember? We demonstrated it at the Hanover Fair."

Bob: "Oh, yes, and your fantastic milling machine … "

Christof: "… with the option of 30-tool-capacity tool changers."

Bob: "Exactly! Oh dear, I've got some machine oil on my hands from one of the vices. Do you think I could wash them somewhere before we go on?"

Christof: "Certainly, the washrooms are next to the entrance of this workshop. Just go back into the corridor and you'll see them on the right. The men's washroom is the first door. Well, the signs will help you anyway."

b Describe the positions of the machines and tools. Use the phrases on page 60.

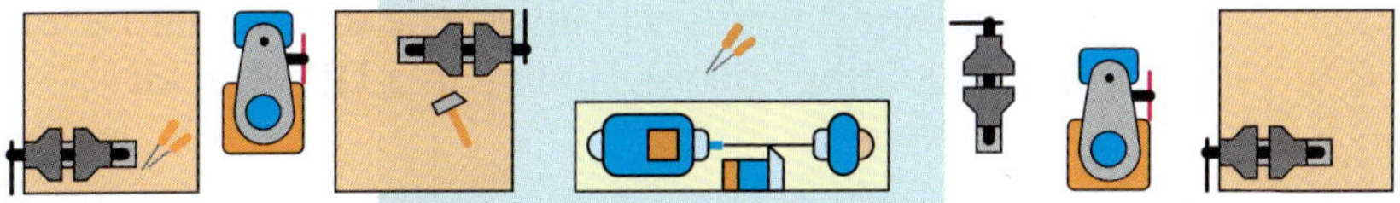

Phrases: **Describing positions**

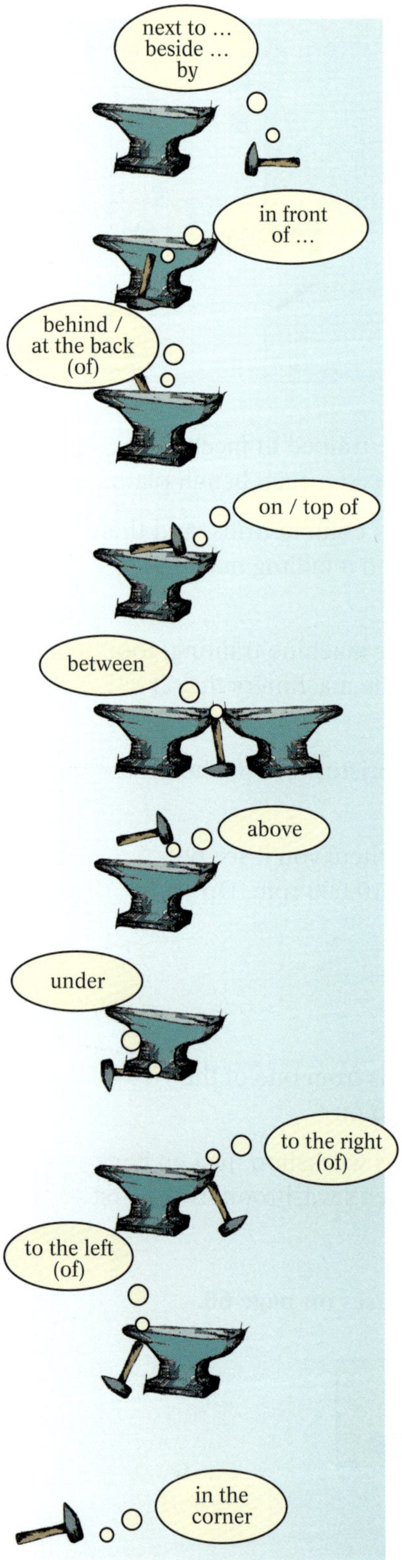

c Complete the following sentences with the words in the box below.

a spindle – a centre lathe – apprentices – a chief executive – skilled workers – the shop floor – milling machine – Industries Fair – manufacturing centre

On the ground floor of a machine tool company you often find ① with all sorts of machinery.

All companies normally have a group of managers with ② as their head.

The world's most famous ③ takes place in Hanover every year.

In the general workshop all ④ receive their first year of practical training.

For working round materials you use ⑤.

The CNC shop is the heart of DEP's ⑥.

All machine tools need ⑦ for performing revolving motions.

After their training the apprentices will be ⑧ who are able to operate a ⑨.

A2 Joining

The joining method you choose always depends on the property of the material to be joined. If you want to join thin pieces of sheet metal, for example, you are most likely to apply riveting techniques. Welding and soldering are often used to repair exhaust pipes and other parts of a car.
High-quality adhesives are used to bond not only plastics and ceramics, but wood and glass as well. Some adhesives can even be used to join metals.
In spite of modern technology, however, a hammer, a screwdriver, a spanner and a wrench – along with bolts, nuts, nails or screws – will always be the standard equipment for joining and assembling in a DIY tool kit.

a Give examples of the joining methods mentioned in the text. What other joining and assembling methods can you think of? Make a list.

b Work in pairs and make up a short dialogue by matching the sentences below.

Example:

Student A: First, I've got to assemble these two base plates of a machine vice.

Student B: Well, you'll need socket screws and nuts.

First, I've got to assemble these two base plates of a machine vice. • And after that I need to tighten the bolts of the machine vice. • Next, I need to weld the two pieces of sheet metal over there. • And before we can take our coffee break, these two flat steel plates must be joined to form a square. • By the way, I'm afraid I broke the handle of the coffee pot in our office. Have you got some adhesive?

I wouldn't weld them, because the material is aluminium. I'd rather rivet them. The box of rivets and the riveting tongs are over there on my workbench. • *Well, you'll need socket screws and nuts.* • I suppose you'll need a spanner and an Allen key, won't you? • Yes, I have. There's a new tube of mixed adhesive in the top drawer of my desk. • Well, now you have to weld, of course. You can use my welding torch, but be careful when you're handling the welding equipment.

c Create dialogues using the following prompts. Make up any details.

Example:

Student A:
1. I want to loosen these screws.
2. I've got to join these two wooden parts.

Student B:
1. You'll need a screwdriver.
2. Why don't you nail them together?

Student A:

I want to … I've got to …	join tighten/loosen bond	two wooden boards. two metal bars. … .

Student B:

Why don't you …	weld … solder … nail (together) … rivet … screw/unscrew …

You'll need …	adhesive. a soldering iron. a hammer. riveting tongs. a screwdriver. a spanner. a welding torch.

B Machine tools and power tools

B1 Machine tools

For cutting processes the centre lathe, the milling machine and the grinding machine are some of the most important machine tools. Although in industry these machine tools are used as CNC (Computer Numeric Control) machines for large-scale manufacturing with high accuracy, the conventional machine tools are still used in training centres.

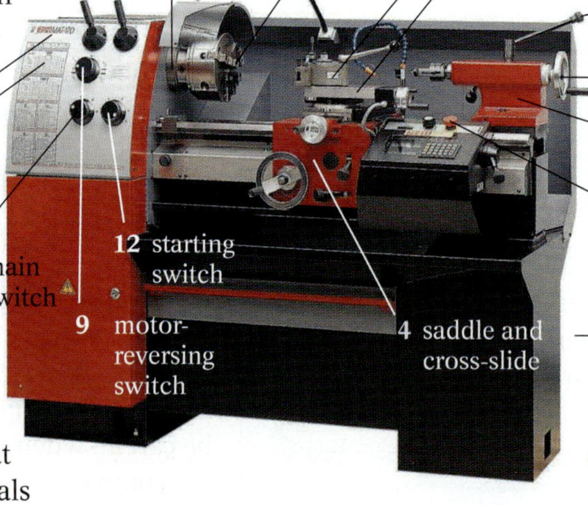

centre lathe

The principal difference between a centre lathe and a milling machine is the shape of the components that are machined. Round materials are manufactured on a centre lathe and flat materials are worked on a milling machine. A grinding machine is used for sharpening tools and for accurate smoothing of surfaces.

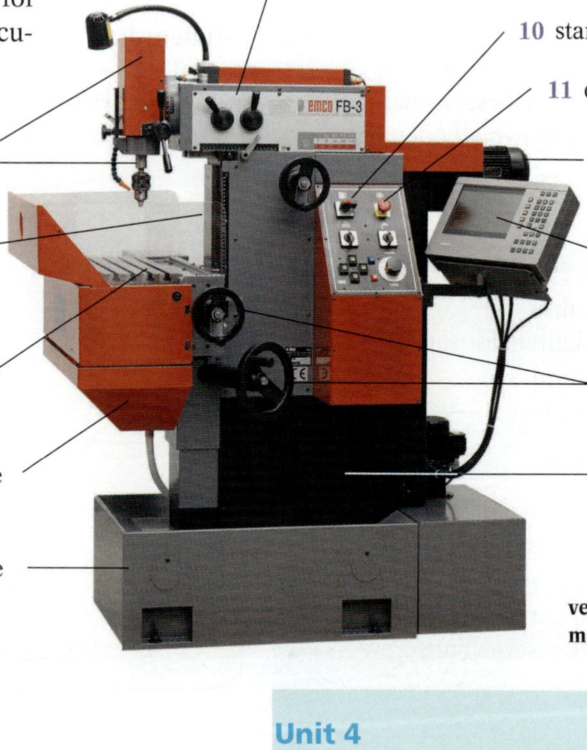

vertical milling machine

The lathe, the milling machine and the grinding machine are metal-cutting machines that have some basic elements in common such as the bed, frame or base, the machine table (grinding machine and milling machine), different clamping systems (e.g. chucks: lathe and milling machine), the main switch and the emergency stop, some slideways, and various gear and control units.

Each machine needs a motor that provides the power to drive the gear and the control units, which transmit the power to the headstock (centre lathe) or the different spindles. The revolution speed of a spindle is set with the speed-selection lever. The way the various slideways of all these machine tools are operated depends on the type of machine, but normally they are used to control the different feeds of either the cutting tool or the workpiece.

Some elements are typical only for special types of machine tools.

A **lathe**, for example, is equipped with a saddle, a headstock, a chuck, and a toolpost. The headstock drives the main spindle, which rotates the workpiece. This rotation can be performed either clockwise or anticlockwise and is controlled with the motor-reversing switch. The saddle and cross-slide feeds the cutting tool which is mounted to the toolpost.

Milling machines need a column, an overarm with the spindle, and the vertical slide for the machine table. The table supports a machine vice to clamp the workpiece in the required position.

Typical elements of a **grinding machine** are a base, a column, a base table to store grinding wheels, the table traverse and a grinding wheel. The grinding process, which is probably the oldest form of metal cutting, uses an abrasive cutting material. The abrasive particles are normally bonded together in the form of a wheel.

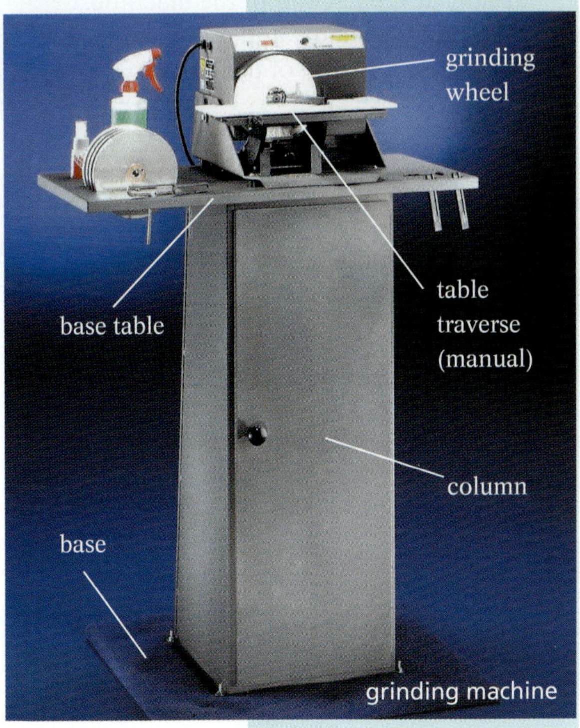

a Match the French names of the machine elements with the corresponding English terms.

base table	affûteuse
bed/frame	arrêt d'urgence
base	arrêt-marche
centre lathe	base
chuck	bâti
column	capot
emergency stop	chariot longitudinal
grinding machine	chariot porte-outils
grinding wheel	colonne
headstock	contre-poupée
knee	coulisse verticale
main switch	disque abrasif
milling machine	fraiseuse
cross-slide	mandrin
tailstock	poupée
toolpost	table
table traverse	table transversale
vertical slide	tour à pointes

b Match the French and English terms above to the type of machine tool they refer to (lathe; milling machine; grinding machine).

B2 Centre lathe

saddle and cross-slide / tailstock

a Work in pairs. Student A asks about the loction of a machine part as shown in the picture on page 62. Student B studies the picture again and answers the question using the following prompts.

Example:

Where exactly is the starting lever? *It's on the bottom right-hand side of the headstock.*

Student A

Where exactly is the ...?

headstock
emergency off
tailstock
main switch
chuck
toolpost

Student B

It's ...

on the nose of the main spindle.
at the top left end of the lathe.
on the top right-hand side of the control panel.
usually at the left end of the machine.
on top of the saddle and cross-slide.
at the top right end of the lathe.

Phrases: Describing the location of machine parts

at the left end of – at the right end of – on top of – at the bottom of – on the nose of – at the rear/back of – at the front of – underneath – on the bottom right-hand side of – on the bottom left-hand side of – on the top right-hand side of – on the top left-hand side of

b Work in groups. One student asks about the function of the tool element, another gives the answer and continues with the next question, while a third or fourth student corrects the answer if necessary.

Example:

A: What is the headstock needed for? – B: It houses the driving gears for rotating the spindle.

Student A

What is the … needed for?

What's the function of … ?

What is a … used for?

- chuck
- *headstock*
- machine bed
- saddle and cross-slide
- toolpost

Student B

Well, it …

- supports the saddle and cross-slide and the toolpost during the operation
- *houses the driving gears for rotating the spindle*
- holds the cutting tool during the turning operation
- holds the workpiece during the whole turning process
- provides the feed for longitudinal turning of the cutting tool

Student C

I'm afraid …

… I don't agree here. It seems to me that the main function of … is …

That's not quite right. It's the ... that houses/provides/supports/holds …

c Explain the function of the parts of the centre lathe in the box on the right with the help of the reading text on the previous pages. Carry on with the work in dialogue groups as you have done before.

Parts of a centre lathe:	
motor	main switch
gear and control units	motor-reversing switch
emergency switch	speed-selection lever

B3 Milling machine

a Name the elements whose locations are described in the following sentences. Look at the picture on page 62 for help.

1. It's at the rear/back of the body.
2. It's at the front of the column.
3. It's on top of the column.
4. It's next to the emergency stop.
5. It's on top of the knee.

overarm

motor

table

vertical slide

start-and-stop switch

overarm (horizontal milling machine)

b Replace the following words and explanations with synonyms from the text on page 63.

1. parts	5. tool
2. counter-clockwise	6. produces
3. toolholder	7. material that is being worked on
4. assembled	8. the rate of advance of a cutting tool

c Match the machine elements with their functions.

knee	A. switches the electricity supply to the machine on/off.
base	B. moves the table towards or away from the cutter.
table handwheel	C. houses the drive motor.
start / stop switch	D. forms a part of the column and supports the tank of the coolant.
column	E. carries the machine table and houses various machine controls.
overarm	F. rotates the milling cutter during the operation.
table	G. provides the headstock and the main spindle.
spindle	H. carries the machine vice that holds the workpiece during the operation.

B4 Grinding machine

- **Complete the following sentences with the names of the components which are listed below.**

the grinding wheel – the table traverse handwheel – the base table – the table – the main switch – the cross-feed handwheel

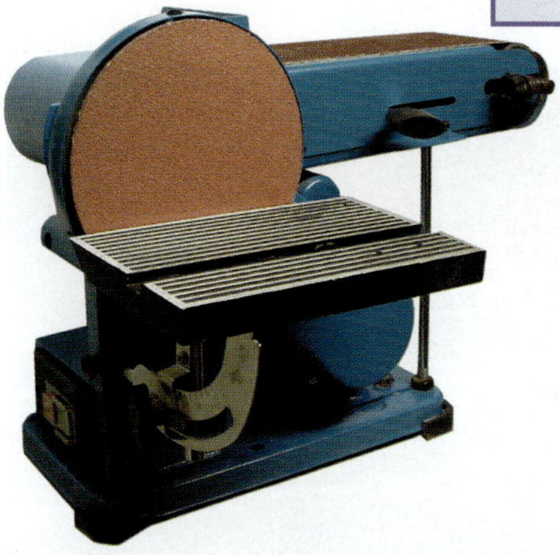

1. ... provides the table cross-travel.
2. ... stores grinding wheels.
3. ... carries the desired workpiece to the grinding wheel.
4. ... controls the electrical supply to the machine (on/off).
5. ... removes the material from the workpiece.
6. ... provides the longitudinal table movement.

B5 Power tools

Power tools have become indispensable. They are used not only in industry but also in many households. DIY stores offer a wide range of power tools such as impact drills, angle grinders, jigsaws or sanders. They all need electric power, either connected or cordless like light-duty drills.

a Name the power tools shown on the Internet page below.

b Match the following product features with the power tools above.

Product features:

- Ideal for sanding large panels
- New battery technology for maximum runtime
- Standard equipment: chuck key, side handle and depth gauge
- 48 external torque settings for precise settings of screws
- Supplied complete with grinding disc and handgrip
- Variable preselection of speed for optimum drilling results
- A compact powerful tool, ideal for general grinding jobs in the garage, workshop at home
- 2-speed power-tool, ideal for DIY, professional, automotive and industrial use
- Reversible drive with adjustable torque settings, screwdriver bits, spanner and adapters
- Compact power tool, ideal for sanding car bodies

c Under which link buttons on the Internet page above would you find the following items?

batteries – screwdrivers – battery-driven impact drills – drill bits – a box for your tools – chainsaws – gloves – band saws – goggles – wrenches

d Make a list of tools you need to do the following jobs. Use a dictionary if necessary.

1. Assembling and setting up bookshelves.
2. Restoring an old chest of drawers.
3. Connecting electric wires.
4. Mounting a nameplate to a wall next to the doorbell.
5. Joining metal bars for a garden fence.
6. Shortening the bars of a metal shelf.
7. Finishing the surface of a repaired wing (mud guard).
8. Deburring the edges of sawed bars.

e Read the following sales dialogue in pairs. It's about a cordless screwdriver drill.

Customer: I'm interested in buying a driver drill. What would you recommend?

Assistant: Well, let me show you our latest model. It's a reversible cordless screwdriver drill with a totally new design. It comes with a two-speed variable trigger switch, and you can set five torque adjustments for superior power.

Customer: Sounds impressive. What about the service life of the battery?

Assistant: This model comes with two batteries and their service life is more than 2 hours of operation. An interesting new feature is the red signal lamp that indicates low charge. For recharging there is this quick charger.

Customer: What do I do with empty batteries that cannot be recharged anymore? You know, I can't dispose of them with the domestic waste.

Assistant: That's no problem. We are happy to take those batteries back and dispose of them in an ecologically friendly way.

Customer: Well, you said something about quick recharging.

Assistant: Yes, it's a 9.6-V DC 1-hour rechargeable battery cartridge and a 1-hour charger with an auto shut-off.

Customer: Great. What about safety?

Assistant: That's another improvement. The drill has a new safety feature that locks out the power when not in use.

Customer: Fine. Does the carrying case have enough space for drill bits?

Assistant: Yes, of course. It has separate compartments for bits, battery cartridge and charger. And the material the case is made of is 100% recyclable.

f List the main features of the screwdriver drill pointed out by the shop assistant.

g **Match the English terms in the product description with the French terms below.**

deux vitesses synchronisées

conception ergonomique et compacte

rotation droite et gauche

pour percer des trous jusqu'à 25 mm de diamètre

puissance 650 watts

boîte de vitesse en métal complètement isolée

embrayage de sécurité

jauge de profondeur intégrée

mandrin sans clé pour serrage rapide

h Work in pairs and translate the following conversation between a customer and an assistant about an impact drill. Use the vocabulary from the previous pages.

Shop assistant A:

Customer B:

Bonjour, monsieur...

Bonjour. Oui, je cherche...

A: ... Est-ce que je peux vous aider ?

B: ... un outil électrique pour percer des trous.

A: C'est pour quelle utilisation ?

B: Je voudrais poser des étagères dans mon nouvel appartement.

A: Est-ce que les murs sont en béton ou en brique ?

B: En béton.

A: Eh bien, dans ce cas, je vous recommande une perceuse à percussion. Voici notre dernier modèle. Il a un moteur puissant de 1000 watt et deux vitesses synchronisées.

B: Est-ce qu'il y a une rotation droite-gauche ?

A: Oui, bien sûr. En plus, l'appareil possède un débrayage de sécurité.

B: Est-ce qu'on peut changer rapidement les outils?

A: Oui, ce nouveau modèle a un mandrin sans clé pour serrer les forets.

B: Est-ce qu'on peut percer des trous de 20 mm de diamètre ?

A: Oui, l'appareil peut percer des trous jusqu'à 24 mm de diamètre. Notez aussi la poignée auxiliaire qui rend l'appareil plus facile à manier.

B: Ah oui, elle y était déjà sur le dernier modèle. Par ailleurs, j'ai vu une publicité où il est indiqué que vous reprenez les vieilles perceuses. Est-ce vrai ?

A: Bien sûr, monsieur.

B: Formidable ! Merci beaucoup.

Example:

Student A: Hello, can I help you?
Student B: Hello. I'm looking for a power tool for drilling.
Student A: What do you need it for?

i Your company wants to present the latest models of their power tools in catalogues, leaflets and on the Internet. **Write product descriptions for different power tools using the phrases in the box, the product features given in the texts before and below, and /or product information from the Internet.**

Product features

- High-speed screwdriver for fast screwing
- Metal gearbox for exact seating of bearings and gears
- Powerful 1000 watt motor
- Quick-release bit holder
- Noiseless clutch
- Variable speed through electronic trigger control
- High-strength belt clip

Phrases: Describing and advertising products

This is our latest /top-of-the-range model
There's an interesting new feature …
This model comes with …
The major benefit of this … is
Another advantage of this model is …
The new … model works better than
This device / component is for …
You can use this tool for (cutting/ grinding/ polishing)
It is very reliable, lightweight and easy to use.
There is an extensive range of original accessories for all our power tools.

j Give operating instructions in the imperative (command) form by putting the following sentences into the correct word order.

Example: drill – clamp – chuck – in
Clamp the drill in the chuck

1. battery – cartridge – put – 1-hour charger – into
2. trigger switch – set – low speed – for
3. required – torque – adjust – this drilling operation – for
4. direction of rotation – select – reversing lever – with
5. discharged – battery – replace – recharged – one – with
6. required – revolution speed – read – this table – off
7. plug of the charger – put – that socket – into – over there
8. rotation – reset – clockwise direction – for

C Safety regulations in a workshop

When working in a workshop with tools and machinery you must always be aware of the possible danger of either breaking or ruining something, or even being injured yourself. It is vital, therefore, to wear the right workshop clothing and to follow the safety instructions when setting up and starting a machine.

Five different types of safety signs in Europe:

I. Prohibition signs: Round signs with a white background and red border and cross-bar mean that something is forbidden.

II. Warning signs: Triangular signs with a yellow background and black border warn of special danger.

III. Mandatory signs: Round signs with a blue background and a white symbol show what protective equipment must be worn.

IV. Emergency signs: Square or rectangular signs with white symbols on a green background indicate safe places, such as emergency exits.

V. Fire safety signs: Square signs with white symbols on a red background indicate location of firefighting equipment, alarm, etc.

a List the following signs under the categories shown above (e.g. Warning signs)

b List the following signs used in English-speaking countries under the same categories (see above).

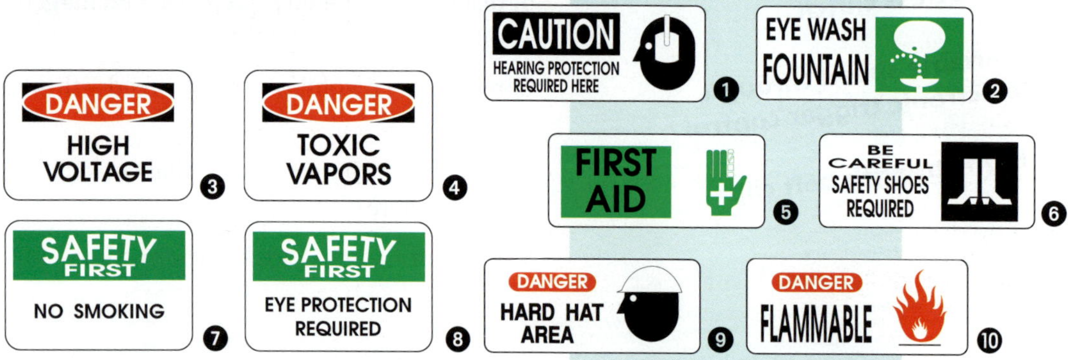

c Match the words to the pictures.

- gloves
- goggles
- hard hat
- ear protection
- fire extinguisher
- guard
- welding mask
- safety boots
- emergency off

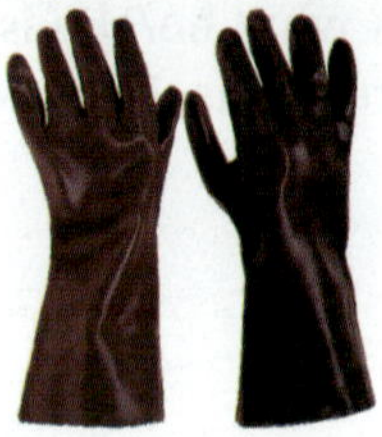

1

2

3

4

5

6

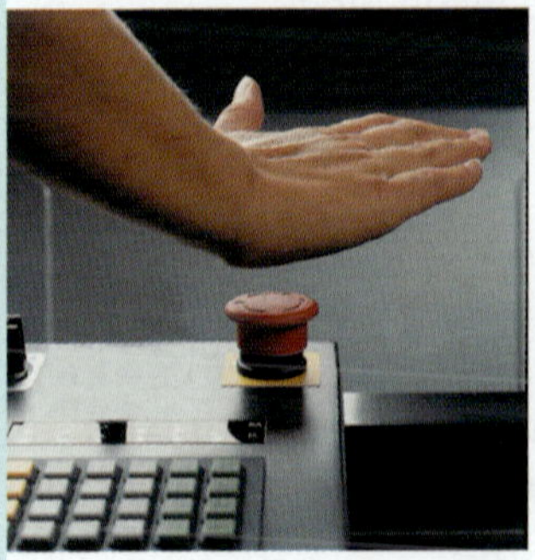

7

8

9

d Complete the safety instructions on the left with the words on the right.

1. Avoid …	a. fire extinguisher
2. When you have long hair, don't forget to wear …	b. guards
3. Wear … when welding.	c. gloves
4. Don't forget to wear your …	d. the mains
5. Position the … of your machine tool correctly.	e. goggles
6. Wear … when you work with sheets of metal or glass.	f. a hair net
7. Switch off … in case of emergency.	g. escape routes and emergency exits are not blocked.
8. Put on your … before you start grinding.	h. loose clothing
9. Take care that …	i. safety boots
10. Take care that the … is always in good working order.	j. safety rules
11. Watch out for … on the floor of the gangways.	k. a mask or a handshield
12. Read all … carefully.	l. slippery oil

e **Work in pairs. Student A says what he/she is going to do. Student B gives appropriate safety instructions. Use the phrases below.**

Student A:

I'm going to ...
I need to ...

stop work for today.
bring in a new column drill.
drill some holes in the plate.
grind the surface of the metal block.
carry heavy metal stock into the milling shop.
weld two bars.
repair the cable of the hammer drill.

Student B:

Use ...
Don't ... (forget to ...)

keep the gangways free.
return all tools after use.
wear safety boots.
wear a mask or handshield.
watch defective insulation.
wear hair net.
wear goggles.

f **Work in pairs and write down safety instructions. Then compare them in your class and list them on the blackboard. Use the following phrases.**

Phrases: Giving safety instructions

Avoid ...
Don't forget to ...
Wear (e.g. gloves) when ...
Position ... correctly.
Switch off ... before you open the casing.
Put on (goggles) when/before ...
Read the instructions carefully before ...
Never ...
Always use protective ...
Make sure ... before ...

D Grammar revision: Questions and relative clauses

See Grammar Overview in the appendix.

I. Questions and interrogative pronouns

- **Ask for the words in bold type. Think of the interrogative pronouns and the correct structure:**

Example:
1. Who entered the reception hall of the company?
2. Who did he meet at the exhibition in Hanover?

1. ***Bob Morris*** *entered the reception hall of the company.*
2. *He met the* ***chief executive of DEP*** *at the exhibition in Hanover.*
3. Mr Watson showed his visitor to **the shop floor of the firm.**
4. The apprentices get some basic training at the machine tools **because they'll be operating them as skilled workers later.**
5. Opposite the centre lathes you can see a **universal milling machine.**
6. The metal-cutting machines will be set up **next month.**
7. These tools can be used in several ways **to join parts.**
8. The motor provides the power **to drive the gears.**
9. The headstock is supported **by the column** in vertical machining centres.
10. The workpieces must be clamped **in a machine vice.**

II. Relative clauses

Remember!
Relative clauses begin with a **relative pronoun**.
When the pronoun is the subject of the verb that follows, use "who" (for people) or "which" (for things).

- **Connect the following sentences using the correct relative pronoun : who or which?**

Example:
Some machine tools are in the general workshop. ***They*** are not used for manufacturing.
Some machine tools which are not used for manufacturing are in the general workshop.

1. The factory often has visitors from abroad. The visitors are required to wear hard hats during the tour.
2. The apprentices are doing their first year of training. They are presently working at a milling machine.
3. The table supports the machine vice. The machine vice clamps the workpiece in the position required.
4. Joe needs some help with his work. He is one of DEP's skilled workers.
5. The lathe tool must be sharpened for the facing operation. It can be used for various turning jobs.
6. Our latest lathe model is equipped with a high-speed spindle. We demonstrated the model at the Hanover Fair.

Unit 5

Troubleshooting, maintenance and warranties

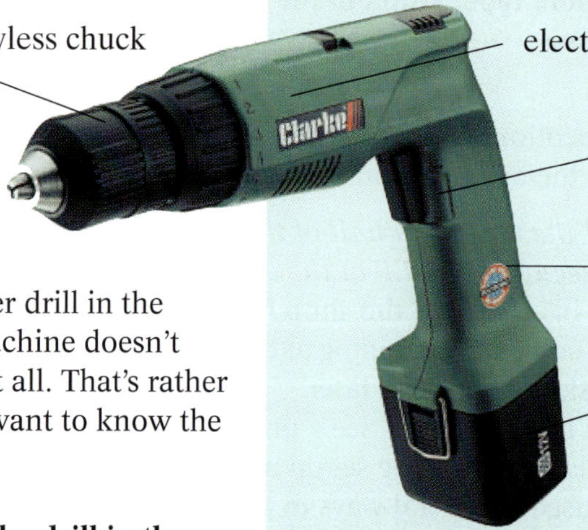

You have made your work plan, you have figured out all the details required for your job, you are ready to start working with a machine like the cordless power drill in the picture on the right, but the machine doesn't work properly or doesn't run at all. That's rather annoying. And of course, you want to know the reason, don't you?

a Find possible reasons why the drill in the picture above might not work by completing the sentences with the words in bubbles.

1. The battery is …
2. The keyless chuck is …
3. The twist drill is …
4. The trigger switch is set …
5. The electric motor is …

damaged | blocked | low | anti-clockwise | loose | broken | blunt

b Work in pairs. Ask questions (Student A) and give answers (Student B) using the following prompts.

Example:

Student A:

Why / electric motor / not running
Why is the electric motor not running?

Student B:

Because / battery / must / (re)charge
Because the battery must be (re)charged.

What / wrong / twist drill?
Why / twist drill / not clamp / properly?
What / wrong / electric motor?
Why / drill / not make / proper hole?

- because / keyless chuck / blocked
- it / blunt / must / and / sharpen
- because / switch / set / anticlockwise
- it / damaged / must / repaired

c Work in pairs and make dialogues using the following pictures and prompts. Prepare your dialogue and look up any missing words if necessary.

Student A:

The ... is out of order. Do you know what's wrong with it?
(Sorry,) the ... doesn't seem to be working (properly), could you ...?
I can't get the ... to work. Could you take a look at ...?

Student B:

Maybe ... is the problem.
Have you checked ...?
Let's have a look at ...
It looks like the ...
motor / handle / housing / plug / switch / chuck / twist drill / battery ...
... is broken / has come loose / has cracked / is jammed / is blocked / is low ...

A Troubleshooting – A column drill

Steve Pearson, an apprentice at DEP, is about to drill holes in the base plates of the machine vice which he has to manufacture. As the column drill he is going to use does not run he asks his master, Dave Crewe, to help him.

- **Put the following dialogue in the correct order. Then listen to the CD and check if you were right.**

Steve Pearson:	Yes, it's not working and I don't know why.
Dave Crewe:	Hmm, the spindle's set fine and the drill's properly clamped in the chuck. But what about the revolution speed?
Steve Pearson:	Well, as the drill is an HSS centre drill I set the speed as high as possible, which is 25,000 revolutions/minute.
Dave Crewe:	What's the problem, Steve? Any trouble with the machine?
Steve Pearson:	All right. And could you please check if the machine is set correctly?
Dave Crewe:	Did you check the main switch?
Steve Pearson:	Will do. And thanks again!
Dave Crewe:	Fine, and you know that after centring you read the required revolution speed from the table here at the front of the machine gear box?
Steve Pearson:	Yes, I know. The diagram in the table tells you the ratio of revolution speed and diameter of the drill.
Dave Crewe:	Well, let's see. Look here, Steve, the safety button is still locked in, you see? To release it, you must turn it a bit to the right and then pull it out, okay?
Steve Pearson:	Yes, of course I did. That was the first check, and I checked the safety button as well.
Dave Crewe:	Yes, you've got it. If you have any more problems, just give me a ring.
Steve Pearson:	Oh yes, you're right, thank you. And the machine vice must be in central position as well?
Dave Crewe:	And think of the machine table. It should be set high enough for you to drill properly into the material.

B DIY Power Tools Manual

Tips on drilling

DIY – The Do It Yourself market has expanded enormously during recent years due to the increasing number of people who prefer to do craftsman's work themselves. Power tools such as drills, hammers, saws, and grinders have become more and more attractive and professional, not to mention easier to use. All you have to do is follow the instructions in the manuals.

Of all these DIY-tools the drills seem to be the most popular. The power drills available have high technical standard and are very versatile.

Which drill for which application?

There are certain points that have to be considered before drilling. You need to know the properties of the material concerned, e.g. wood, metal, stone or concrete. The type of material determines what kind of drill bit should be used and what drilling speed should be selected.

For drilling **wood** a twist drill bit with a brad point, which drills holes with small to medium diameters, is recommended. A twist drill bit called an HSS (high-speed steel) drill should be used for drilling **metal**. When the material is **stone** or **concrete** a tungsten-carbide drill bit is used. Quite often the DIY worker prefers a multi-purpose drill that can cope with most common material properties.

Which drilling speed?

Which drilling speed should be selected? You can find relevant information on the packaging or in the operating instructions. However, there is a basic rule that says the smaller the drill bit diameter, the higher the speed, and the larger the drill bit diameter, the lower the speed. A second rule states: the softer the material to be drilled, the higher the speed, and the harder the material to be drilled, the lower the speed.

a Match the English terms from the text above with their French translations.

drill bit
saw
concrete
multi-purpose
diameter
craftsman
do-it-yourself

scie **foret** **polyvalent** **béton** **bricolage** **artisan** **diamètre**

b Complete the following text using *finally, after that, then, first, now.*

How do we go about drilling holes with our brand new high-power drill? ① we have to study the instruction manual carefully. ② we need to decide what kind of material is to be drilled. ③ we should determine the diameter of the hole to be drilled. ④ we can decide which drill bit should be used. ⑤ we should check the required drilling speed.

C Maintenance and warranty
C1 Maintenance of a power drill

MAINTENANCE

- The ventilation slots of the machine must be kept clear at all times to prevent overheating.
- If the machine is mainly used for impact drilling, collected dust should be removed regularly from the chuck. To remove the dust hold the machine vertically with the chuck facing down, and open and close the chuck completely. The collected dust will fall from the chuck.
- The regular use of cleaner spray for the clamping jaws and the clamping jaw borings is recommended.
- Use only original STAR TOOLS accessories and spare parts. Should components need to be replaced which have not been described, contact one of our STAR TOOLS service agents (see our list of guarantee/service addresses).
- The range of accessories with part numbers is shown in our catalogue. We also recommend our CD-ROM with 1500 exploded views and 100 000 data sets.

DRILLING OPERATIONS

Drilling in masonry

1) Take the drill out of the hole from time to time to remove dust.
2) Switch to impact drilling for concrete, hard bricks, stone, hard cement, and marble, except when drilling the surface of marble. Use drill bits with soldered hard metal (= carbide) tips.

Drilling in metal:

1) Centre-punch the point you have marked for drilling.
2) Always hold sheet metal firmly in a vice.
3) Lay a block of wood under thin metal to prevent it from distorting.
4) Use HSS spiral drill bits.
5) Before drilling large-diameter holes, first drill a small pilot hole.
6) For lubrication use oil for steel, turpentine or paraffin for aluminium.
7) Don't lubricate when drilling brass, copper, or cast iron, but take the drill out of the hole frequently to cool it off.

Screwing:

1) Use screwdriver bits of the appropriate size and shape.
2) Pre-drill into hardwood, or for screws of large diameter or countersunk screws.
3) If you work with softwood, suitable screws can be driven in without pre-drilling.
4) If you operate anticlockwise under high strain the chuck might become loose. It's always better to use the bit reception to loosen screws in anticlockwise operation.

a **Look up all the words and phrases from the text above that fit into the following categories and write them down: 1. cleanliness 2. care 3. spare parts**

Work in groups and compare results.

Example:

1. *cleanliness:* ventilation slots must be kept clear
2. *care:* hold the machine vertically with the chuck facing down
3. *spare parts:* use only original parts

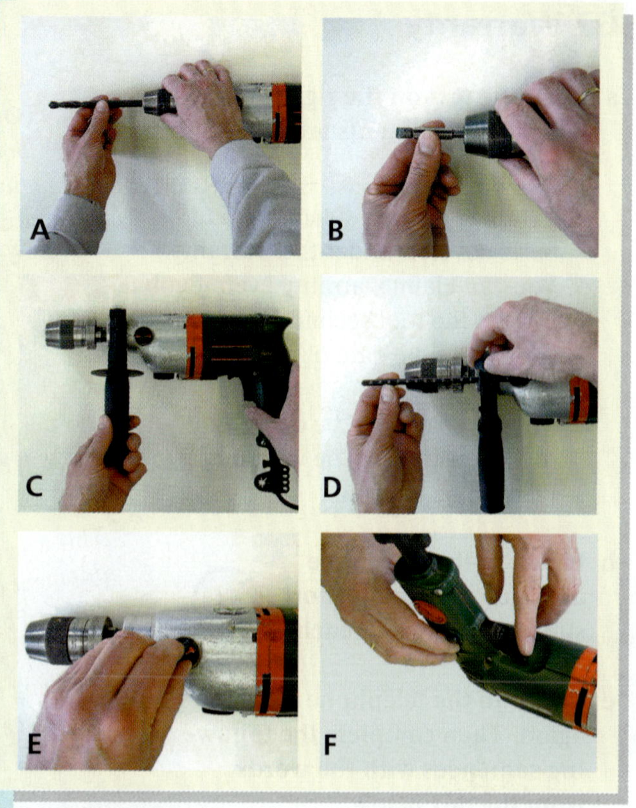

b Match the following instructions with the illustrations.

Instructions: 1. Selecting forward/reverse drive 2. Positioning the auxiliary handle 3. Changing the keyless drill 4. Inserting tools 5. Setting the depth gauge 6. Selecting impact drilling

c Give recommendations and tips for maintenance and drilling operations. Student A: Ask one or two of the following questions. Student B: Find suitable answers from the options below.

Student A (Questions)

1. *When should the collected dust be removed from the chuck?*
2. What has to be done to remove the dust completely?
3. How can the clamping jaws and the clamping jaw borings be maintained?
4. What do you recommend for finding accessories and spare parts?
5. What sort of drills should be used for masonry jobs?
6. Why is it better to lay a block of wood under thin metal?
7. What sort of drills must be used for metal work?
8. What lubricants are used for drilling brass, copper, or cast iron?
9. Can this tool be used for screwing?
10. Can this tool be set anticlockwise for loosening screws?

Student B (Answers)

- Use HSS spiral drill bits.
- *When the machine is used for impact drilling.*
- They must be cleaned regularly with a cleaner spray.
- Yes, of course, but use screwdriver bits of the appropriate size and shape.
- Check our catalogue or CD-ROM.
- Use percussion carbide-tipped masonry drill bits and don't forget to switch to impact drilling before you start working.
- No lubricants are needed, but the drill should be taken out of the hole frequently to cool off.
- Yes, certainly, but in anticlockwise operation the bit reception should be used.
- The machine must be held with the chuck facing down vertically, and the chuck has to be opened and closed completely.
- This way the metal won't distort.

Example: 1. A: When should the collected dust be removed from the chuck?
B: When the machine is used for impact drilling.

C2 Warranty

a Read the text on the right and say if the following statements are true or false.

1. TOOLPRO gives a 3-year warranty.
2. TOOLPRO will replace any faulty parts covered under the warranty.
3. You can claim warranty even if you bought the tool second-hand.
4. A proof of the date of purchase is a must.
5. Minor repairs can be carried out by the customer. The warranty will not be affected.

TOOLPRO LIMITED ONE-YEAR WARRANTY

TOOLPRO guarantees its Power Tools for a period of one year from the date of original purchase. Should you have any problems with one of our Power Tools purchased in the United States, we will either replace or repair any part or parts of the product and accessories covered under this warranty.

This warranty is extended only to the original purchaser.

A purchase receipt or other proof of date of purchase will be required before warranty performance is rendered.

This warranty does not cover damage which occurs in shipment or failures which are caused by products not supplied by **TOOLPRO**, or failures which result from normal wear, accident, misuse, mishandling, modification, using the product with an improper accessory or service by anyone other than an authorized **TOOLPRO** service center.

For repair or replacement, return the complete tool or accessory, transportation prepaid, to your nearest **TOOLPRO** Service Center.

b Listen to the telephone dialogue and give a short summary of it in French.

c Listen to the telephone dialogue again. Then complete the following sentences with the words from the box below.

servicing – defect – power drill – warranty – reversing switch

1. The problem with the (1) is that the (2) won't work.
2. The repair service is free of charge because the drill is covered by a (3).
3. The tool's (4) has to be repaired at an authorized service centre.
4. With regular maintenance the power tools shouldn't need any (5).

d Work with a partner in a role play simulating a telephone dialogue between a power tool manufacturer and a customer.
Use the phrases from the table on page 83.

Student A:	Student B:
I see. When did you buy the tool?	Oh, fine. Thank you for your help. Goodbye.
Yes, of course. We give a 1-year warranty on all our products.	It's the reversing switch which is sticking.
You're welcome. Goodbye.	Ah, that's fine. What do I have to do then?
Hello, this is STAR TOOLS. Can I help you?	About three months ago. Can we claim the warranty?
Just send the tool to an authorised service center. You'll find a list of our service centers at the back of your manual.	*Hello, this is Bill Morris from DEP.* We've got some trouble with your power drill STAR 600 SP.
What seems to be the problem?	

Example:

A: "Hello, this is STAR TOOLS. ... ?"

B: "Hello, this is Bill Morris from DEP ..."

Safety instructions

Before starting to operate the power tool the students are asked to pay attention to the safety instructions in the attached leaflet.

- **Read the following safety instructions and find the missing words in the word square on the right.**

- Follow the accident regulations.
- Appliances used at many different locations including open air must be connected 1 a current surge preventing switch.
- During work on material containing asbestos or stonework containing crystalline silicic acid, dust will arise that is 2 to the health.
- Always wear 3 when using the machine. It is recommended to wear 4, sturdy non-slipping shoes and an apron.
- Do not 5 sawdust and splinters while the machine is running.
- Do not pierce the motor housing because this could damage the double insulation. 6 adhesives instead.
- Always switch 7 before carrying out any work on the machine.
- Only plug in when the machine is switched off.
- Keep 8 lead clear from the working range of the machine.
- Always lead the cable behind you.
- Always use the auxiliary handle.
- Do not use diamond core 9 in impact drilling.
- When drilling in walls, ceilings or floors, take care to 10 electric cables and gas or water pipes.

A	S	Q	V	D	K	P	S
B	A	O	L	R	F	X	H
G	V	I	A	I	R	U	A
G	O	G	G	L	E	S	R
L	I	K	K	L	M	E	M
O	D	L	M	S	O	F	F
V	O	J	A	M	V	R	U
E	S	M	I	L	E	N	L
S	J	U	N	P	L	U	G
N	M	L	S	M	P	M	D

D Grammar revision: Reported speech

See Grammar Overview in the appendix.

- **Work in pairs. Give instructions, recommendations and orders in the following sentences. Use reported speech.**

Examples:

Bill Morris: "**I've got** some trouble with your power drill STAR 600 SP." (told the service engineer) Bill Morris told the service engineer that he **had** some trouble with the power drill STAR 600 SP.

Chris to Matt: "Please hand me the box with the drill bits." (asked)
Chris **asked** Matt **to hand** him the box with the drill bits.

Dave: "**I will** give him the drill bits **tomorrow**." (said)
Dave said (that) he **would** give im the drill bits **the following day**.

1. Instructor to Mike: "Open and close the chuck completely when cleaning dust out of it." (told)
2. Matt to Phil and Sarah: "Don't forget to clean the jaws and jaw borings with cleaner spray."(reminded)
3. Instructor to trainee: "Use percussion carbide-tipped drill bits for masonry drilling." (advised)
4. Instructor again: "And don't forget to switch to percussion drilling before you start working." (he also told)
5. Kate to Mark: "Use HSS spiral drill bits for metal work." (reminded)
6. Trainee to colleague: "When drilling aluminium use turpentine or paraffin as lubricants." (told)
7. Instructor to trainee: "Use screwdriver bits of the appropriate size and shape." (told)
8. Instructor to trainee again: "Pre-drill into hardwood or for countersunk screws." (also advised)
9. "Have you set the switch anticlockwise?" (Bill asked if Steve)
10. "I'll meet the service engineer at the entrance." (Dave said)
11. "The drill will be repaired tomorrow." (Steve promised)
12. "I've been waiting for 2 hours!" (The instructor complained)
13. "Did you check the insulation of the mains yesterday?" (The instructor asked the trainee)
14. "The machine isn't working and I don't know why." (Dave said)
15. "Did you clamp the twist drill properly?" (Dave asked Steve)
16. "Why doesn't the drill make a proper hole?" (Dave wanted to know)
17. "I released the safety button." (Dave said)
18. "The component has not yet been finished." (The trainee told the instructor)

Unit 6

Electrical engineering

No matter where you go you will see electricity at work. Bulbs change electricity into light, electric motors change electricity into motion, electric heaters change electricity into heat. Many devices and household appliances around you are powered by electricity which is either stored in batteries or supplied by a wall socket.

Dishwasher (1 hour / day)
Refrigerator (24 hours / day)
TV (4 hours / day)
Vacuum cleaner (1 hour / week)
Mobile phone (stand-by 14 hours / day)

kWh/year: 432 20 642 38 292

a **Try to match the different devices above to their energy consumption.**

b **Listen to the recording and check your suggestions.**

c **Calculate the annual costs for the devices in the table on the next page. Use the following formula for your calculation.**

INFO-BOX: $\frac{\text{wattage x hours used per day}}{1{,}000}$ = daily kilowatt-hour (kWh) consumption
1 kilowatt = 1,000 watts

Multiply this by the number of days you use the appliance during the year for the annual consumption. If you want a general estimate of how much electricity your home appliances consume, you can usually find the wattage of most appliances stamped on the bottom or back of the appliance, or on its 'name plate' or in the manual.

device	time in use	consumption / hour
hairdryer	15 minutes / day	1,200 watts
printer (stand-by)	7 hours / day	4 watts
printer (in use)	30 minutes / day	16 watts
electric heater (portable)	10 minutes / day	2,000 watts
10 electric bulbs	12 hours / day	60 watts per bulb

Example:

clock radio: 5 watts/hour
in use: 24 hours/day
calculation:

$$\frac{5 \times 24}{1,000} \text{ kWh} = 0.12 \text{ kWh}$$

in use for 365 days: 0.12 kWh x 365 days = 43.8 kWh/year
utility rate: 8.5 cents/kWh
annual costs : 43.8 kWh x 8.5 cents/kWh = 372.3 cents = € 3.72

d Compare any electrical devices as follows.
Form sentences by using the following phrases:

The energy consumption of ... is as low as ...

A TV consumes more / less energy than ...

The energy consumption of ... is higher / lower than ...

Example:

- An electric bulb consumes less energy than a vacuum cleaner.
- A water heater consumes more energy than any other device in the household.

A Some basic facts about electricity

A1 General information

Electricity can be defined as a flow of energy which is produced by the movement of electrons. Electrons are parts of atoms which can be found everywhere, as everything is made up of atoms. When electrons are on the move they produce an electric charge, so electricity can be produced by moving electrons. The flow of electrons is known as **current**, which is measured in amperes (amps).

We distinguish between two types of electricity:

- **Static electricity** is electricity that does not move. You sometimes get a slight electric shock from touching a metal door knob, because the static electricity that builds up in your body suddenly flows from your hand to the door knob. Lightning is static electricity suddenly discharging between clouds and the ground. Static electricity builds up with friction when two different materials rub together, that is, when electrons are moved from one place to another, e.g. by rubbing a plastic comb with a piece of cloth.

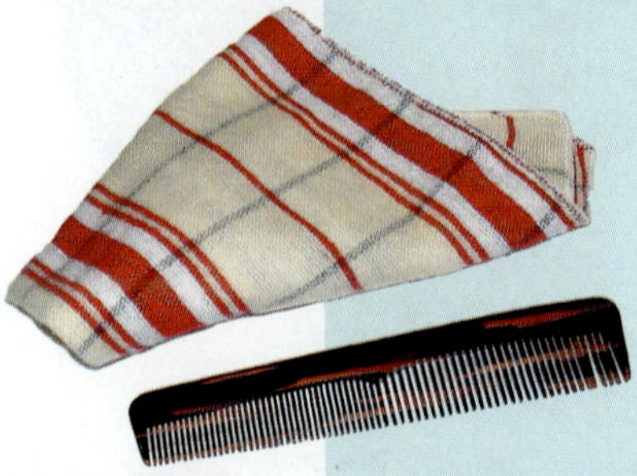

- **Current electricity** is where electrons are moved from one place to another as a result of a chemical reaction (e.g. in a battery). Current electricity can be generated in different ways. The most common of these are explained in the following description:

A2 Batteries (chemical energy)

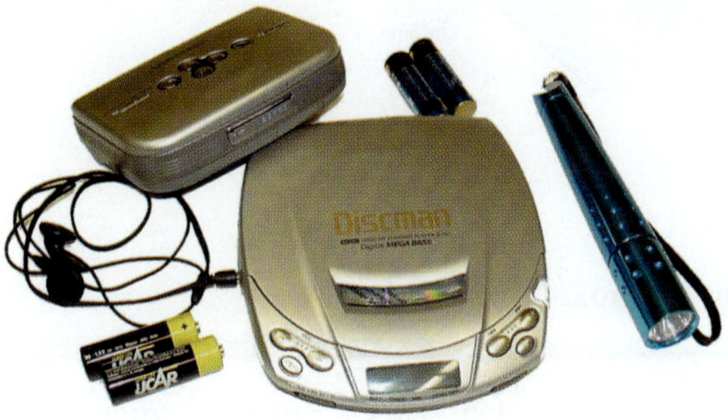

In order to make electrons move, a source of energy is necessary. Chemical energy can be such a source of power. When a rod of carbon and a rod of zinc are immersed in a solution of ammonium chloride, a current of about 1.5 volts flows through a wire connecting the two rods. The electrons will always flow from the negative pole to the positive pole. In our example the zinc rod, which is the negative pole, will slowly dissolve in the ammonium chloride. The flow of electricity will stop when the zinc rod is completely dissolved. The batteries used in cars and torches usually apply this technology.

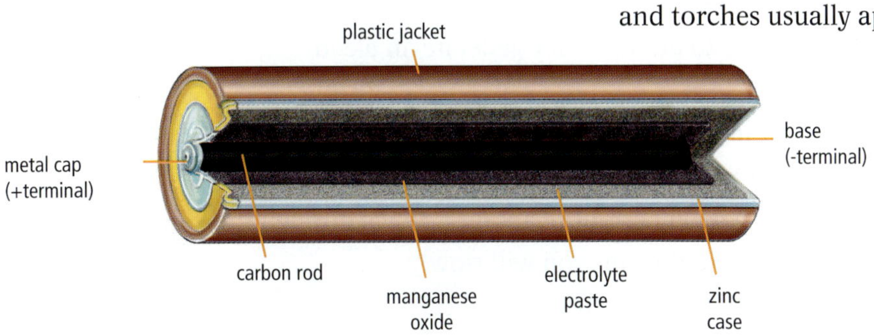

(zinc-carbon cell)

A3 Generators (mechanical energy)

Mechanical energy is produced with the help of magnetism. The forces of a magnet come from two points near its end: the north and the south pole. Poles of the same kind repel each other, poles of opposite kinds attract each other. Every current has a magnetic field around it. If a current flows through a coil or wire, the coil acts like a magnet. When a current flows through the coil it becomes magnetised. The pull of a magnet makes the coil spin round. Generators in power stations use this principle to produce electricity.
A generator converts mechanical energy into electrical energy by employing electromagnetic induction. A conductor is moved through a magnetic field, directing the current produced to an external circuit. The conductor can be a simple coil of wire which rotates between the poles of a magnet. When the coil passes through the magnetic field in both directions, alternating current (AC) is produced. These generators are called alternators. A commutator is used when direct current (DC) is needed, that is, when the current flows in only one direction.

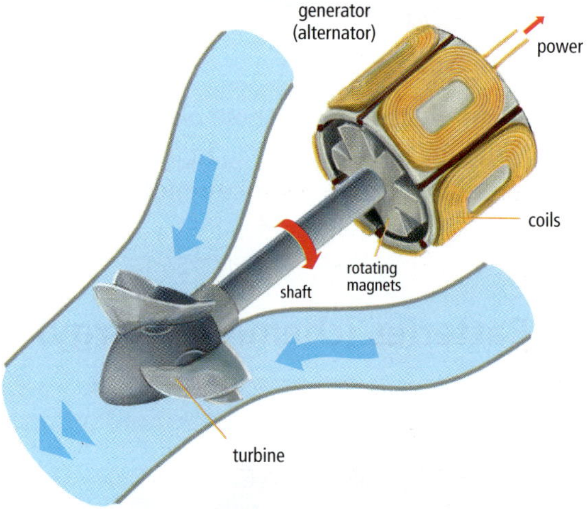

a Describe the difference between static electricity and current electricity.

b Which charges attract and which repel each other?

c How does a generator work?

d Complete the following sentences with words from the box.

ammonium chloride – atoms – applied – generated – repel – 1.5 volts – current – dissolve

1. When electrons move from one place to another they generate an electric … .
2. Like charges always … each other.
3. Electricity can be … chemically or mechanically.
4. In order to produce an electric current in a battery a solution of … is needed.
5. When you connect the rods with a wire, a current of about … flows through the wire.
6. During this process the negative pole of the zinc rod will slowly … .
7. This technology is usually … in the batteries of torches and cars.
8. Everything is made up of … .

A4 Electrostatic induction

The effect of electrostatic induction can easily be tested by the following experiments:

- **Find any missing words in the following texts in the word square below.**

1. You can use a plastic spoon to make a stream of water from the tap flow sideways.

– Rub the plastic spoon on your clothes.
– Move the spoon slowly towards the stream of water from the tap.

Effect: The direction of the stream of water from the ① flows sideways.
Explanation: The negative charges on the spoon ② the negative charges in the water to the far side and ③ the positive charges to the near side. The spoon induces these charged regions in the water, so the effect is called electrostatic induction.

2. Several experiments with blown-up balloons that are charged by friction show the effects of electrostatic induction.

A	R	J	W	F	L	X	O
N	E	U	T	R	A	L	P
V	P	M	C	I	T	A	P
P	E	P	H	C	T	U	O
T	L	E	A	T	R	M	S
F	Y	R	R	I	A	M	I
H	A	N	G	O	C	L	T
M	O	V	E	N	T	O	E

– **Rub a balloon on your ④ and then let go of it.**
What happens? The balloon sticks to the jumper because rubbing the two together gives each one an ⑤ charge.

– **Rub a balloon on your jumper and ⑥ the balloon slowly towards small pieces of paper.** What happens? The balloon that has been charged by ⑦ will attract small pieces of paper as a result of the opposite charges.

– **Take two threads and attach them to two balloons. Hang them side by side before and after charging them by friction. What happens?**
Before being charged by friction the two balloons are both ⑧ – they ⑨ next to each other – no effect can be seen; after inducing them they repel each other because they are both negatively charged.

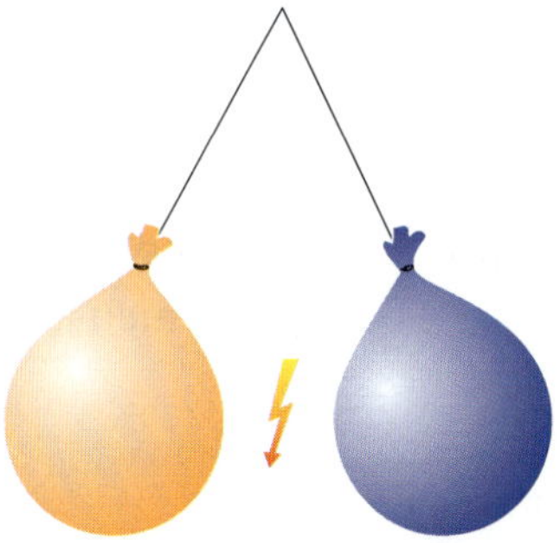

A5 Direct Current (DC) and Alternating Current (AC)

The electric current generated by a battery is **direct current**, where the electrons only flow in one direction, that is, from the negative (cathode) to the positive pole (anode).
The mains supply, however, is not direct but **alternating current**. In this case the electrons move to and fro 50 times a second. The frequency of the alternating current is measured in **Hz** (hertz).
The terminals of the supply repeatedly change from positive to negative and vice versa, which makes no difference to the customer at the end of the line. A bulb will light up when the current flows in either direction.

a Translate the above text into French.

b Match the blue words or phrases with synonyms from the text above and write them down.

1. A direct current is produced in a battery.
2. In an alternating current the electrons move back and forth 50 times a second.
3. In this case the terminals of the supply change again and again from positive to negative.
4. This does not affect the user at the end of the line.
5. The bulb will light up regardless of the direction of the current flow.

c Match the following terms with the explanation.

◆ circuit	Electric current that changes direction continuously.
◆ voltage	The path for an electric current.
◆ atts	A substance that allows the passage of electrons.
◆ amps	The unit of measure of an electric charge.
◆ direct current	Electric current flowing only in one direction.
◆ resistor	A material that does not conduct electricity.
◆ coulomb	When a battery is connected to a circuit, it produces an electric current. This current is measured in ?
◆ electric circuit	The unit of measure of electrical force.
◆ insulator	A ? restricts the flow of the current.
◆ current	The resistance is measured in ?
◆ ohms	The amount of flow of electricity is called ?
◆ conductor	Power is measured in ?
◆ alternating current	A continuous path of conductors and other electrical components along which an electric current can flow.

A6 Tools for electricians

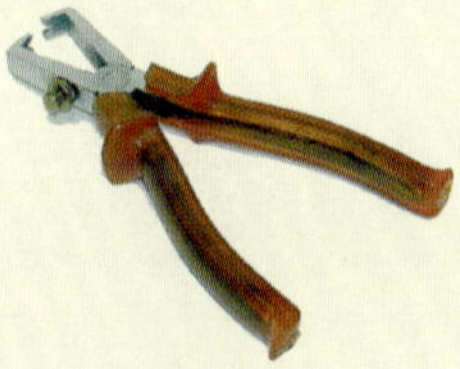

pince à dénuder

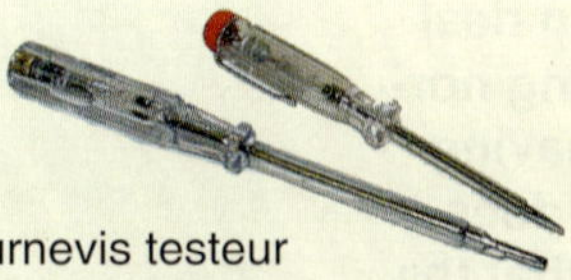

tournevis testeur

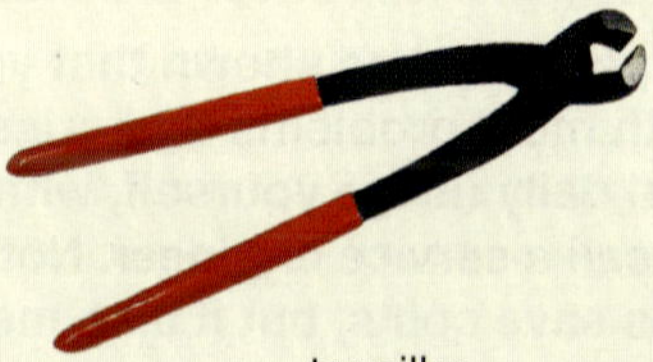

tenailles

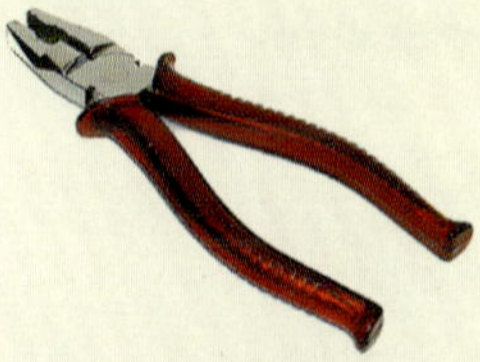

pince universelle

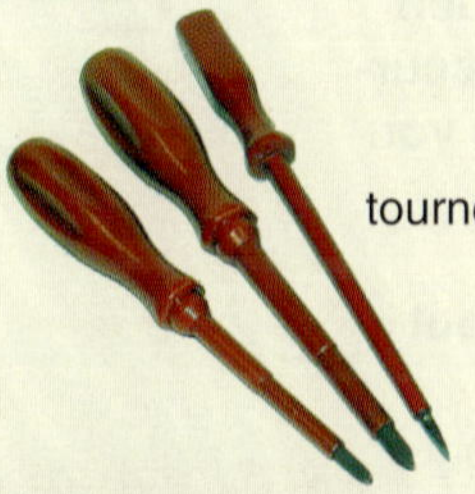

tournevis

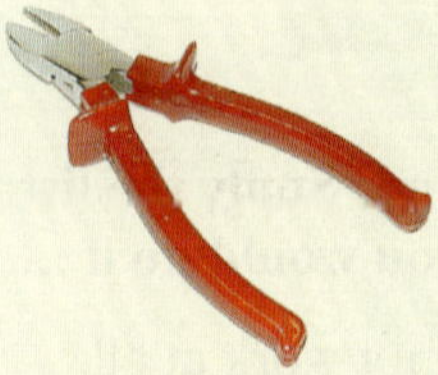

pince coupante

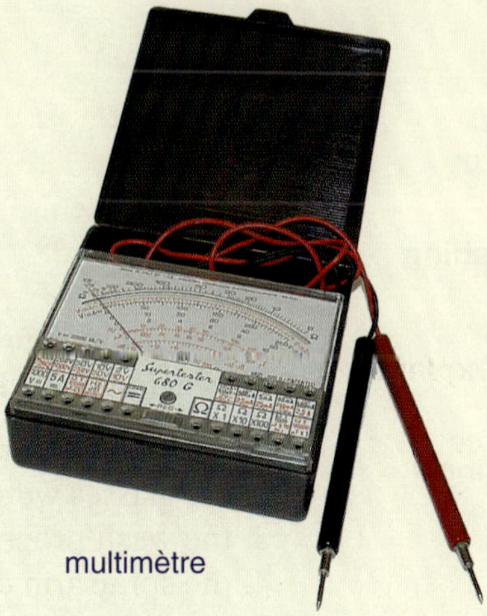

multimètre

1. combination pliers
2. wire strippers
3. screwdrivers
4. diagonal cutting nipper
5. multimeter
6. pincers
7. neon screwdriver (mains tester)

a Match the English terms in the box above with the pictures.

b What are the tools used for? Give examples.

Example:
Wire strippers are used for cutting or stripping wires.

Verbs: cut – saw – check / test – tighten or loosen – hold – bend – hammer – knock in – drive in – screw – turn – solder – chisel – strip – unscrew – put in – pull / take out

A7 Troubleshooting

Dealing with minor problems yourself

Experience has shown that you can deal with most problems that arise during normal daily usage yourself, without having to call a service engineer. Not only does this save costs, but it also means that the appliance is available for use again much sooner. The following list of common occurrences and their remedies should help you identify the causes of most problems:

When the dishwasher is switched on but does not start up:

- Fault with fuse in mains electricity supply.
- Plug not inserted in wall socket.
- Appliance door not closed properly.
- Water tap not turned on.

The spray arm rotates with difficulty:

- Spray arm is blocked by small items or remains of food.

Appliance stops suddenly while washing is taking place:

- Electricity supply has been cut off.
- Water supply has been interrupted.

Glasses have a milky appearance:

- Not enough detergent was added to input compartment.
- No salt has been put into appliance.
- Water softener is set too low.

- **Solving minor problems: Study the list above and say what you would do if ...**

1. the dishwasher did not work at all.
2. the glasses were not cleaned properly.
3. the wash programme stopped suddenly.
4. the spray arm did not rotate.
5. there was no water in the dishwasher.
6. if salt and cleaner were added properly but dishes were still milky.

(First of all) I would check ...

Then I'd try to ...

Next I'd see if ...

Finally I'd ...

Example:
If the dishwasher didn't work at all, I would check the fuses. Then I'd ...

A8 Electricity and safety

Electricity is very dangerous because an accident can be fatal. Therefore, we must be aware of all possible dangers before work on electrical apparatus is carried out. Most accidents happen from contact with live conductors or faulty equipment.

a When using electrical devices, you must follow certain safety rules. Put the words in each box in the correct order.

wet hands with electrical equipment do not touch

an electrical device do remove the casing not from

any cables damaged do use not

dispose of don't the dust bin in batteries empty

touch electric wires never live

wet hands never bulbs replace with

working on electrical equipment before the mains switch off

hair dryer use a never bath tub in the

use the flex/cable never to pull the socket out of plug

outdoors indoor electrical appliances don't use

Example:

Do not touch electrical equipment with wet hands!

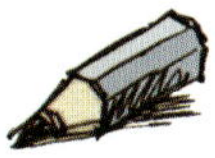

b First measures after an electrical accident

Coming into contact with an electric current can be life-threatening. The extent to which humans can be injured by electricity depends above all on the level of current, but also on the path of the current through the body and how long the electric shock lasts.

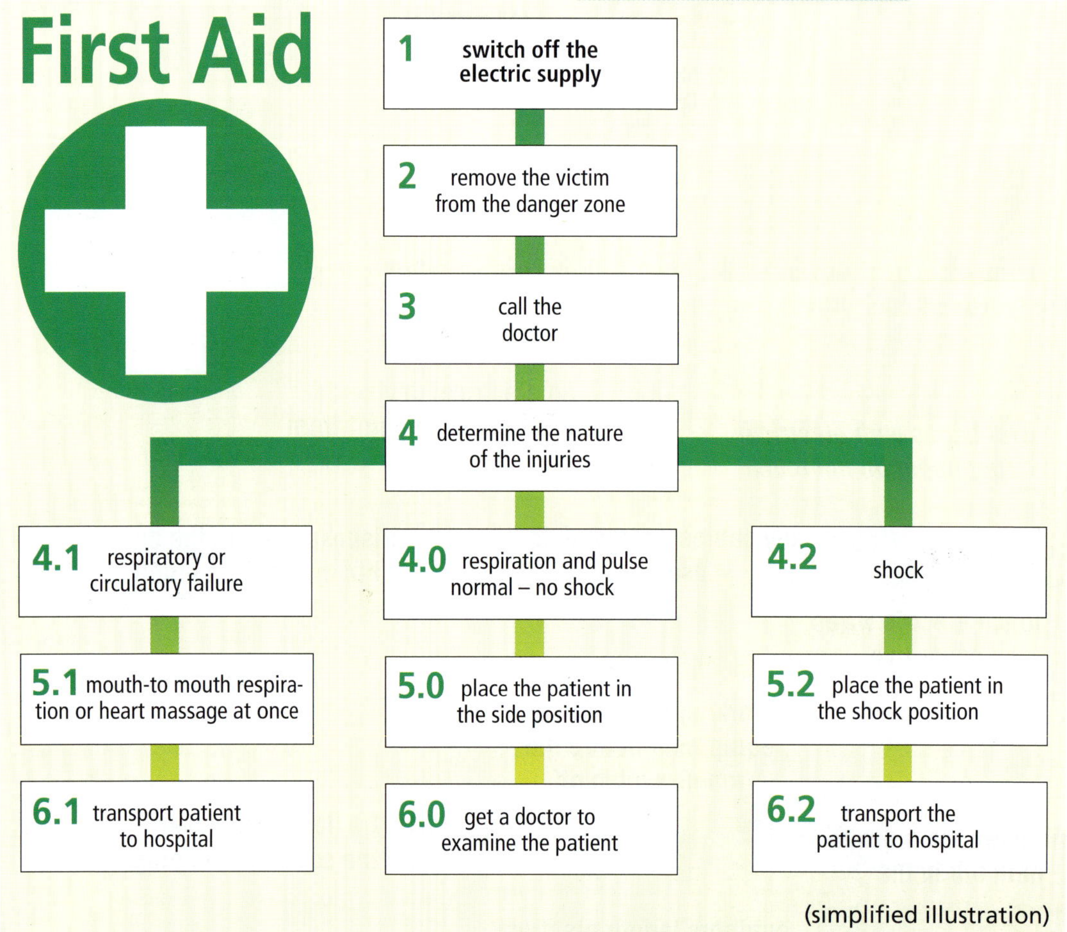

(simplified illustration)

c What would you do before the doctor arrived?

Example
In case of respiratory failure, I would try mouth-to-mouth respiration right away.
If this did not help, I would start heart massage.

In case of …, I would place the patient in the … . Then I would …

B Electricity applied

B1 The car battery

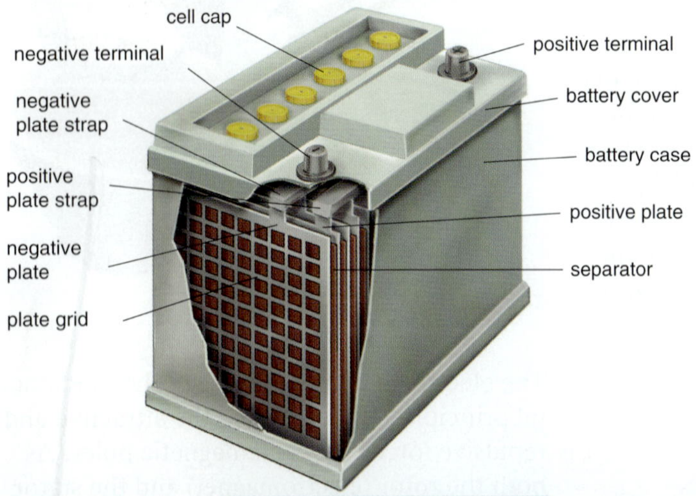

A car battery is a storage battery because it can store the electrical energy produced by the generator. The battery's voltage depends on the number of cells linked together. Cells can be connected in series (that is, positive to negative) to produce a higher voltage. Since each cell provides 2 volts, six of these series-connected electrode stacks form a 12-volt battery, which is needed for a normal car. Alternatively, cells can be connected in parallel, that is, positive to positive, negative to negative. In the latter case the battery produces only 2 volts but the amperage is about six times higher than that of a single cell. A car battery contains plates of lead dioxide and lead metal which are immersed in sulphuric acid and which are connected in series. When the battery produces current (**discharging process**), both kinds of plate change to lead sulphate. As soon as the engine is running again, the current flows back into the battery (**recharging process**) and the chemical reaction is reversed. During the recharging process the lead sulphate formed during the discharging process becomes lead dioxide, lead, and sulphuric acid again. This is the principle of all rechargeable batteries.

a Answer the following questions on the text.

1. What is the effect when you connect cells in a parallel way?
2. What is the total voltage of parallel-connected cells?
3. How does a car battery work?
4. How is a car battery normally recharged?

b Describe how the voltage of a battery can be increased.

c Find the corresponding nouns, verbs, adjectives, adverbs in the text above.

1. numerous (noun)
2. dependence (verb)
3. connection (verb)
4. alternation (adverb)
5. container (verb)
6. sulphur (adjective)
7. chemistry (adjective)
8. recharge (adjective)

B2 The electric motor

The electric motor is clean and silent and starts instantly. It can be large enough to drive the world's fastest trains or small enough to run a watch on your wrist. Its source of energy is supplied along wires from an external power supply or from batteries within the device. The motor can run on direct current which is provided by a battery or alternating current supplied by the mains. While a generator converts mechanical energy into electrical energy, the electric motor is a device for converting electrical energy into mechanical energy.

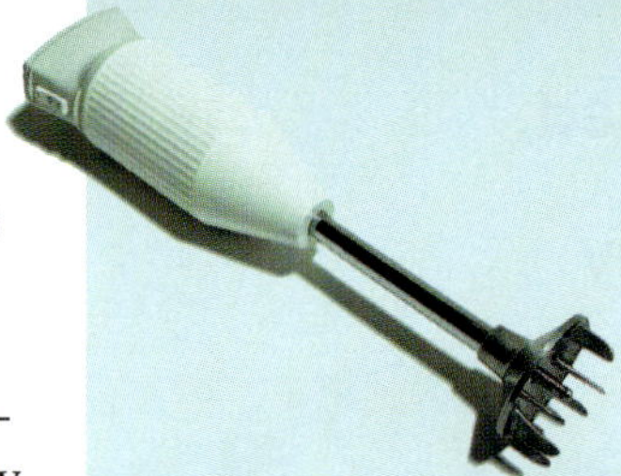

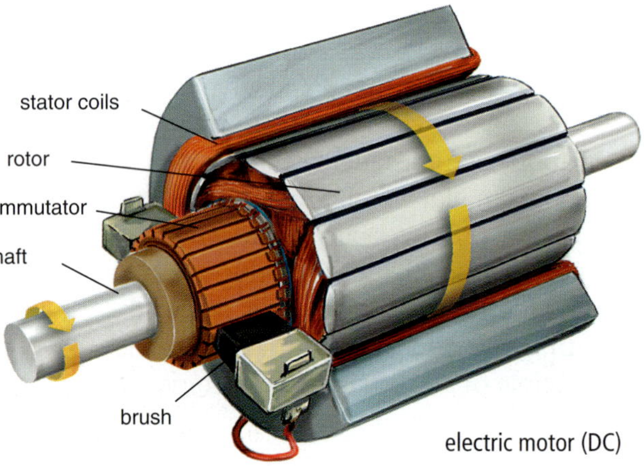

electric motor (DC)

The electric motor makes use of the fundamental principles of magnetism, the attractive and repulsive forces between magnetic poles. As both the rotor (electromagnet) and the stator (stationary magnetic field) are magnetic, the rotor turns as its north pole is attracted by the stator's south pole. In a direct current motor (see illustration on the left) a **commutator** switches the poles and the rotor continues turning in the same direction. The rotor consists of several wire coils wound round an iron core which is connected to a commutator. When an electric current passes through the rotor coil, the coil rotates and the current is supplied through brushes at the commutator.

a Make a list of electrical devices that are run by an electric motor.

Example:
electric can opener, electric screwdriver

b Describe the pros and cons of electrically driven cars. Use the phrases from the box below.

Phrases: describing pros and cons:

I think (that) ...	Nevertheless ...
I believe (that) ...	The first disadvantage is ...
In my opinion, ...	In addition to that ...
From my point of view, ...	Furthermore ...
On the one hand, ...	Consequently ...
On the other hand, there are certain disadvantages in connection with, ...	To sum up ...
	In conclusion ...

c Summarise the principle of the electric motor in your own words. Give information about its

- environmental aspects
- possible applications
- sources of energy
- technology (direct current motor)

d Translate the following technical terms into English. If necessary, read the text on page 96 again to find the words.

1. source externe d'énergie
2. courant continu
3. secteur
4. moteur à courant continu
5. bobine
6. commutateur
7. courant électrique
8. force attractive

C New technologies: Hydrogen and the fuel cell

C1 The fuel of the 21st century: Hydrogen

The idea of using hydrogen as a fuel is not a new one, but interest in it has grown in recent years. Unlike most other fuels, hydrogen cannot be produced directly by digging a mine or drilling a well. Although it is the universe's most abundant element, hydrogen is present in the atmosphere only in concentrations of less than one part per million (ppm). Hydrogen for large-scale use must therefore be extracted chemically from hydrogen-rich materials such as natural gas, water, coal, or plant matter. Production techniques which are now being used include steam reforming of natural gas and electrolysis of water. A number of other technologies are being studied, including several that produce hydrogen from water or biomass.

There are two principal ways in which hydrogen can be used as a fuel. Cars or buses can burn hydrogen in an internal combustion engine (ICE), similar to a conventional gasoline engine, or they are powered by fuel cells running on hydrogen. Although pollutant emissions from hydrogen-fuelled vehicles are minimal, the fuel cell is generally considered preferable in the long term because it allows higher efficiency and a longer range on the same amount of fuel. Furthermore, fuel-cell technology allows the manufacture of 'zero-emission vehicles' (ZEVs).

Up to now the only major non-experimental use of hydrogen as a fuel has been in the space programme, for propulsion and as an on-board source of electricity. Recently, however, the first fuel-cell-driven vehicles have been developed and the first fuel-cell-driven urban bus is expected to be on the market shortly. One of the main obstacles to the introducton of fuel-cell cars on the mass market is the lack of infrastructure, e.g. refuelling stations, service and maintenance.

Fuel cells are said to have the greatest potential regarding alternative drive systems. They are extremely quiet and energy efficient and – even more important – have either very low emissions or none at all.

INFO-BOX:

What is a fuel cell?

A fuel cell is an electricity-producing device running on hydrogen. The hydrogen reacts with oxygen from the air, thus generating voltage between two electrodes. The reaction takes place in an electrolyte solution.

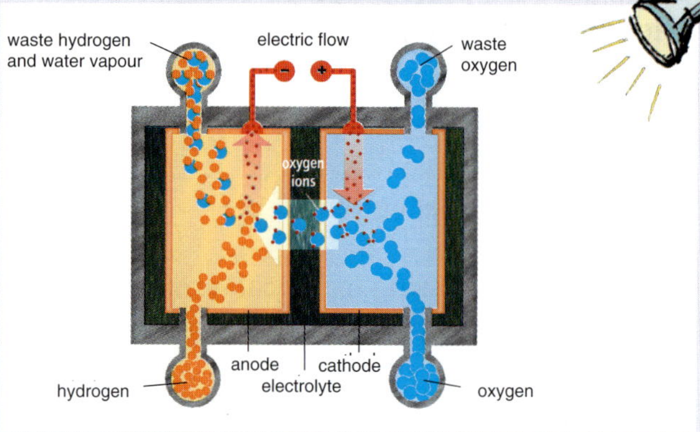

a Give a short summary of the text on page 97 in French.

b You are a marketing assistant. Working in teams of 3-4 students, prepare a presentation on hydrogen as the fuel of the 21st century including the following points. Use the phrases in the box below to help you.

1. Describe the different methods of generating hydrogen.
2. What is hydrogen used for in the space programme?
3. How can hydrogen be used to power vehicles?
4. What is the basic principle of the fuel cell?
5. What are the main advantages of fuel-cell technology over traditional drive systems?
6. What are the main problems regarding mass production of fuel cell cars
   a) according to the text?
   b) according to your knowledge?
7. What are "zero-emission vehicles"?

Phrases: Presentations

I've organized the presentation in (3, 4 ...) parts.
First of all I'll give you a short ...
Firstly ...
Secondly ...
Moving on to ...
After that ...
Then ...
And finally ...

Let's start with some general information on ...
That's all I have to say about ...
Now, let's move on to the second point.
If there are any questions, I'll be happy to answer them.

Useful words: therefore, however, clearly, in the same way, anyway, alternatively, in addition, to sum up, to conclude, in other words, in particular

c Find nouns in the text on page 97 corresponding to these words:

1. technical
2. electrolytic
3. electrify
4. emit
5. combust
6. efficient

d Find synonyms in the text on page 97 for the following words.

1. not long ago
2. in contrast to
3. common
4. existing
5. main
6. traditional
7. more desirable, better
8. moreover
9. city, inner-city
10. absence

C2 Power Engines' fuel cell system

a Michel Arnaud's first week at Power Engines starts with a general introduction by John Smiley.
Listen to the dialogue between Michel and John Smiley and answer the following questions on the text.

1. What are the disadvantages as far as the efficiency of an ICE is concerned?
2. Why is the fuel cell more efficient than a battery?
3. Which other positive side effect does the fuel cell have in contrast to the ICE?
4. What are the advantages of fuel-cell cars over battery-driven cars?

b The fuel cell is the solution to our environmental problems. Do you agree?

c What are the latest developments in fuel-cell technology? Check the Internet or any other up-to-date source.

D Grammar Revision: Passive Voice

See Grammar Overview in the appendix.

a Complete the text using passive constructions for the verbs in brackets.

Example:
In a fuel cell electricity (1. generate) in a direct way. In a fuel cell electricity is *generated* in a direct way.

After hydrogen (1. store) in a tank, it can be used instantly. In the past a bulky fuel tank (2. need) – but technology has improved enormously and hydrogen storage tanks are much smaller now. With technology becoming more and more sophisticated, the efficiency of the fuel cell will be further increased in the near future. Before the catalytic converter (3. add) to the ICEs, there was a high concentration of harmful emissions. Nowadays cars (4. mostly equip) with catalytic converters – but the problems of pollution haven't been solved yet. The answer to this problem can only be found in a completely new engine technology. The best solution is offered by so-called ZEVs (zeroemission vehicles), which are presently (5. develop) by some car companies.

b Describing a process: Change the following instructions into passive voice.

Components: battery, electric wire, switch, socket for a lamp, bulb
Tools needed: neon screwdriver, combination pliers, multipurpose tool

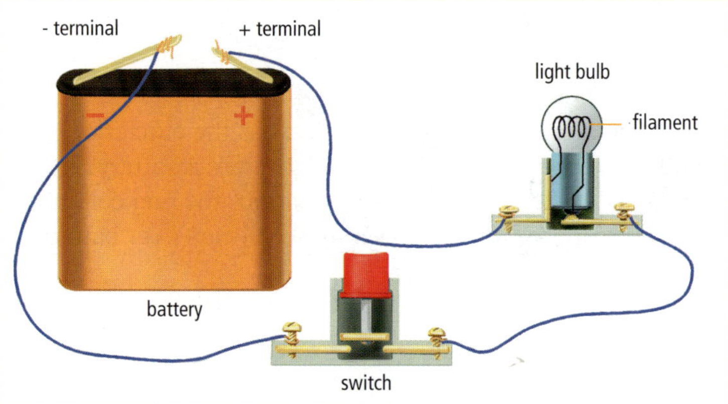

1. Place all components in front of you.
2. Use combination pliers to cut the electric wire into three pieces of equal length.
3. Strip the insulation from the end of the wires.
4. Connect two pieces of the electric wire to the lamp socket.
5. Connect the switch at one end of the electric wire.
6. Connect the other spare wire to the switch as shown in the picture.
7. Put the bulb into the socket.
8. Make sure you turn off the switch.
9. Fix the open end of the wires to the battery.
10. Switch on the power supply.

Example:

1. All components are placed in front of you.
2. The combination pliers are used to cut …

Unit 7

Properties of materials and quality standards (ISO)

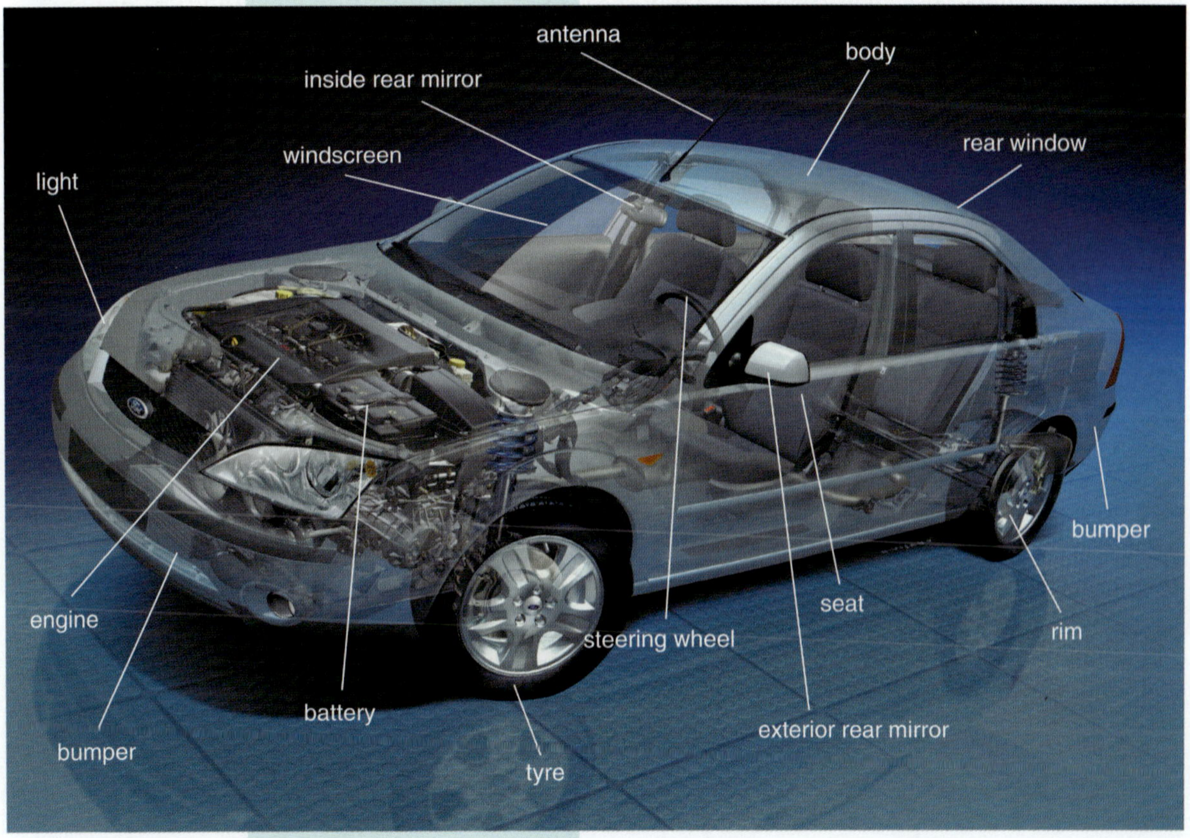

- **Which of the following materials are used for which parts of a modern car? Give reasons.**

copper

lead

aluminium

ceramics

iron

wood

plastics

steel

fibreglass

A Describing products

A1 Which material for which design?

a Three students of design are discussing a project they have to do for their school. They have to design new furniture.

Listen to their conversation and make a list of all the materials they talk about.

b Listen again, check your list and add the properties mentioned to the materials. Discuss the reasons given by the students for their choice of materials.

The following words and phrases may be helpful.

Phrases: Describing materials and products

It is a strong / sturdy / rigid / soft / light / shiny material.
It looks good / modern / state-of-the-art …
It is easy / difficult to work with …
It is something new / classic / timeless / old-fashioned

c Work in groups: Design bookends, a CD-rack, or a desk tidy. Make sure that the language you speak in your group is English. Put the design on a transparency and label the parts; explain and describe it to your class. Use the following words and phrases.

Words and phrases: Presenting a new design

Phrases: our idea is / was • our design consists of • it is constructed of • it consists of / it is made of • the advantage of the material we used is • the (only) disadvantage is … • its measurements are • we decided to use … because … • it can be used for …

Words: length • diameter • width • height • weight • right angle • rectangular • circle • circular • square • glued together • welded together • screwed together • made out of one piece

d Describe the overhead projector in your classroom. If you don't have one, try to find one in the next room.

As a rule for any description, go from the general to the specific. In other words, start with the main body of the appliance, its size and functions, then move on to the specific parts and their functions. Prepare a list with the key words of your presentation; practice it, so that you do not have to check your list too often. Keep eye contact with your audience instead.
The following list of words might be useful:

words to use: overhead projector

to project • transparency • projection screen • housing • fan • transformer • projection lamp • surface • glass plate • lead • switch • mirror • reflector • lens • handwheel • adjust • focus

e Shapes:

When describing objects you need to know the names for geometrical shapes and figures.

Match the following shapes with the corresponding terms.

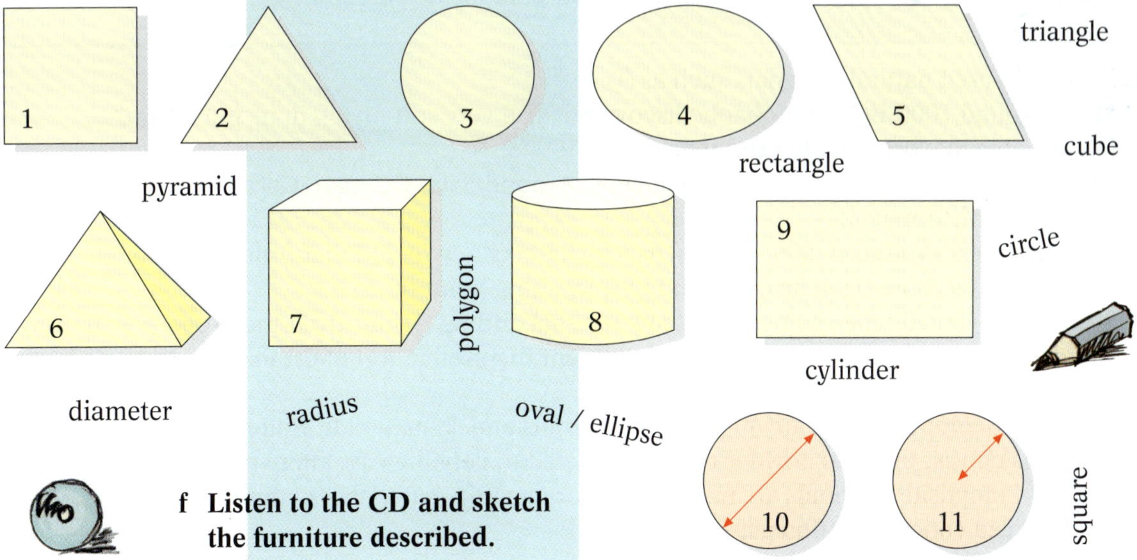

f Listen to the CD and sketch the furniture described.

B Engineering materials

B1 Materials – an overview

One aspect of engineering can be described as the art of finding the most suitable material(s) for a certain purpose or project. That is why an engineer needs to have a very thorough knowledge of the properties of the materials he / she intends to use.

The following chart gives a short overview of different materials.

Metals				Sintered-powder metals	Non-metals			Composite materials
ferrous metals		non-ferrous metals		metal powders subjected to high pressure, then heated to below the melting point	natural materials	artificial materials		a compound of two or more different materials, e.g. fibreglass reinforced synthetics
steels	cast iron	heavy metals	light metals		e.g. wood, stone	ceramic material e.g. glass, porcelain	plastics	
		e.g. lead	e.g. aluminium					

a Into which of the above groups would you put the following materials?

leather – gold – silver – mild steel – titanium

b Complete the sentences below using the words from the box.

synthetics – pressure – temperatures – carbon – furnace – ore – lead – wood or stone – ferrous – alloy – copper – porcelain – sintered – composite – different

- First, man used natural materials such as ①.
- Then, around 7000 BC, he probably discovered ②, a very soft metal, in its pure form. He was able to work it with a stone hammer.
- Copper and lead belong to the non-③ metals. Around 3500 BC man was able to melt copper out of copper ④.
- ⑤ metals were produced later, as soon as a ⑥ had been invented that produced enough heat to melt iron ore. Iron has a much higher melting point than ⑦.
- Steel was invented when someone discovered that adding ⑧ improved the properties of iron.
- Hand in hand with the processing of metals went the production of ceramics such as glass and ⑨. Here too, high ⑩ were needed.
- Materials of modern times are ⑪-powder materials and ⑫ materials. Sintered-powder materials consist of metal powder that is subjected to very high ⑬; afterwards they are heated. Composite materials are made of two or more ⑭ materials. A common example is reinforced concrete or fibreglass-reinforced ⑮.

c List all materials mentioned in the text on page 104, starting with those man used earliest and ending with those most recently developed.

d Check the chart on page 104 and list examples for the different materials used in your trade – for tools as well as for products.

e Work in groups: Choose one or two of the materials from the chart on page 104 and find out about their use and importance, where they are found or produced and what their properties are. You may use encyclopedias or the Internet. Materials about which you can expect to find a lot of information are steel, aluminium, sintered-powder metals, or plastics. Present your results to the class either on handouts or on transparencies, or using computer software you are familiar with.

B2 Alloys

Pure metals are often weak and corrode easily, but their properties can be improved by mixing them with other metals or non-metals. This mixture of two or more metals is called an alloy. Pure copper, for example, is very soft, but if you mix it with tin you can produce a much harder metal. The addition of tin lowers the melting point of the metal and makes it easier to cast. This alloy is called bronze. Another copper alloy is brass, which uses zinc instead of tin.

There are alloys from almost all metals. Steel, for example, is an alloy of iron and carbon. It is harder than pure iron and much more resistant to corrosion.
If you add a certain amount of silicon to iron and very little carbon you get steel with high electrical resistance.
Alloys are much more widely used than pure metals. Aluminium alloys are employed in the construction of aeroplanes, automobiles, cooking pots or building materials. Pure aluminium parts corrode too easily, so they would be unsuitable for use in aeroplanes and buildings. Pure gold and silver are often too soft to be used in jewellery or in dental applications, e.g. for inlays or crowns, so goldsmiths and dentists use alloys instead.

a Give a definition of an alloy based on the text above.

b Why are alloys important in engineering? Give examples.

c The text mentions only a few copper, iron and aluminium alloys. Use your technology handbook, your encyclopedia or the Internet to find more alloys and their use. Give a short report to your class.

B3 Steel

Steel is a metal of many uses which can be found almost everywhere. Simple household objects such as cans for food and drink, the paper clips on your desk, or most sophisticated constructions of buildings and bridges are made of steel. It is said that steel constitutes up to 70% of the weight of a car.

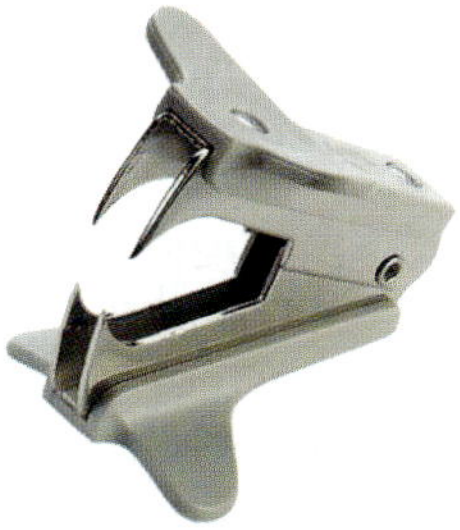

- **What is steel?**
  Complete the following statements on steel, using the words in the box on the right.

> iron – contain – amounts – alloy steel – stainless – containing – producers – sheet steel – alloy – electrical

1. Steel is an ___ of iron and contains up to 1.7 % carbon.
2. Low-carbon steel is an alloy of ___ and contains about 0.2% carbon.
3. High-carbon steels ___ 1% or more carbon.
4. Apart from carbon, steel sometimes consists also of other elements such as manganese, phosphorus or silicon. This kind of steel is called ___ .
5. Steels with small ___ of other metals are called carbon steels.
6. ___ steel is an alloy steel containing at least 11% chromium.
7. For cutting tools the hardest steel is used, ___ up to 20% tungsten.
8. Japan, the USA and Russia are the world's biggest steel ___ .
9. High-quality steel is produced in an ___ furnace.
10. You can work steel by hammering it while it is still hot, or you can press it between rollers to produce ___.

B4 Ceramics

When talking about ceramics most people think of pottery, art and the like. Few people know that ceramics have also become an indispensable alternative to many other materials used in modern technology. Today ceramics are used for a variety of high-tech applications such as engine components, teeth, joints or cutting tools. These applications of ceramics are often referred to as advanced or technical ceramics.

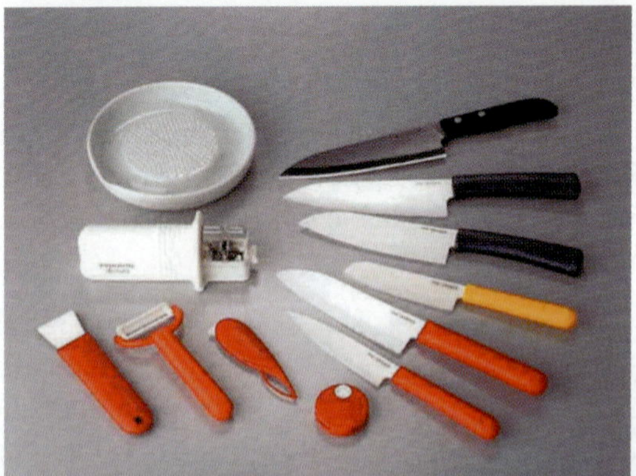
Household utensils

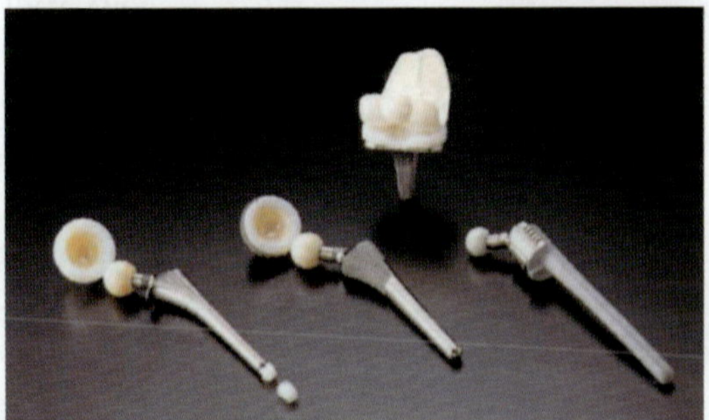
Artificial hip joints

ceramic:
inorganic, non-metallic material made from firing clay, also from metal oxides, or compounds of metallic elements, carbon, nitrogen, or sulphur. Hard and brittle material used in household and industry.

a Make a list of ceramic products and parts. Use a dictionary if necessary.

Example: sinks, valves, ball bearings

b Look at the list of properties in the box on the right. Which of the properties are not typical for ceramics?

Properties:

- high electrical conductivity
- enduring utility
- elasticity
- mechanical strength
- chemical durability (for example against corrosion by acids, bases, salts)
- hardness, resistance against wear
- low electrical conductivity
- brittleness
- softness

c Give reasons why ceramics are used for certain products. Make sentences.

Example: Ceramics are used for the manufacture of insulators because of their low electrical conductivity.

d Gather more up-to-date material on ceramics in industry, from either an encyclopedia, a magazine, technology books or the Internet. Focus on what they are used for and on what advantages they have over other materials. Using this information, prepare a presentation to your class (either on a transparency or as a computer or Internet presentation): Your presentation should contain examples of products, the most important properties of ceramics, and your own conclusions.

I have chosen the following example(s) to show you (how) … Here the use of ceramics has clear advantages over … No other material has the same resistance to wear/heat … When comparing ceramics with …

B5 Plastics

Don't we live in a plastic world? We use plastic buckets in the kitchen, give plastic toys to our children, use plastic casings for computers, printers and many electrical appliances in our households, and we even drive in plastic cars. Almost everything around us contains plastic components, and plastics have become a key material in the modern world.

Though they are generally very susceptible to heat and chemicals, plastics are often used in industry (e.g. the automobile, aerospace and packaging industries), where they are called engineering plastics. Unlike general-purpose plastics that are used in every household, engineering plastics compare well with metals and other materials regarding their strength, heat and wear resistance.

Plastics are not only of comparable strength, they are also lighter and cheaper than many other materials, and unlike metals and wood, plastics do not rust or rot, which can, on the other hand, be a problem when it comes to disposing of plastic products. Many of these products cannot be burned, as they produce poisonous fumes. So the disposal of plastics is still a big waste problem.

Plastics are divided into two groups, thermoplastics and thermosetting plastics, depending on their reaction to heat. Thermoplastics can always be softened by heating and hardened by cooling, i.e. they can be re-moulded. Thermosetting plastics harden permanently after being heated once, so they cannot be reshaped. The difference in reaction to heat between thermoplastics and thermosetting plastics is due to the chemical structure of the plastics. Thermoplastics can be heated and cooled, and consequently softened and hardened, again and again. Consequently, thermoplastics can be reused almost indefinitely. Thermosetting plastics cannot be melted down again. This explains why they are often used in the manufacturing of heat-resistant products, as they can be heated to temperatures of 260° C without melting.

a Make a list of major areas where plastics are used today.

Example: car industry

b Where would thermoplastics and where would thermosetting plastics best be used? Think of the different properties and give examples.

Plastics have replaced ...

... is made of ...

For heat-resistant products ...

... is a suitable material for ...

c Complete the following text with words from the box.

raw – engineering – granulate – advantages – coal – heat – moulded

Plastics belong to the most important ① materials nowadays. They are man-made, and they have numerous ②. First of all, they can be produced fairly easily from ③ materials that are in most cases plentiful: ④, oil, lime, salt and water. The ingredients are bought in sacks in the form of ⑤. The processing is also quite simple. By the application of either ⑥ or pressure, sometimes supported by chemical reactors, the granulate is molten and ⑦ into practically any form or shape desired. The results are perfectly moulded objects that are lightweight, as strong as required and durable.

d Make a list of the advantages and disadvantages of plastics mentioned above and on page 108. Can you find more with the help of your technology handbooks or other sources of information?

e Listen to the recording and write down the names of the materials mentioned. Make sure the spelling is correct.

f The chemical names of the compounds in exercise **e** are certainly confusing, but some of them have more common names.

Match the following terms with their synonyms.

nylon

Teflon

polymethylmethacrylate (PMMA)

acrylics

polytetrafluoroethylene

polyamide

g Name some products made of acrylics, nylon or Teflon.

B6 Mechanical properties of materials

a Match the terms with the descriptions.

– This property allows materials to be stretched without losing their strength. In contrast to elastic materials these materials keep their new shape.

Ductility

– The property of materials that can be worked into a new shape by hammering, rolling or pressing.

Elasticity

Durability

– The resistance of materials to abrasion, cutting, etc. which is required for the moving parts of machines or the cutting edges of tools in a lathe.

- The resistance of materials to corrosion, e.g. gold and platinum.

– The capability of a material to return to its original shape and size after being deformed.

Hardness

Malleability

b What are the adjectives from the following nouns?

Example: ductility – ductile

ductility – malleability – hardness – durability

c What are the nouns of the following adjectices?

brittle – conductive – elastic

d Describe the properties of the following materials using words you have just learned. Try to make at least two full sentences.

Choose from the following materials:
glass, copper, iron, stainless steel, aluminium.

Example: Ceramic material is very brittle, but also very hard and durable. It has no electrical conductivity.

e What properties are needed in materials that are used to produce the following products and parts?

cutting edges – screwdrivers – antennas – valves – ball bearings – springs – door locks – aeroplanes

B7 Elements and their physical properties

a Find out the melting points of carbon, cobalt and tungsten. See table 1 in the appendix for help.

b Give these melting points in Fahrenheit. To convert Fahrenheit into centigrade subtract 32, multiply by 5, and divide by 9. To convert centigrade into Fahrenheit, do the reverse: multiply by 9, divide by 5, and add 32. See table 2 in the appendix.

C Measurement and quality control (ISO)

C1 Measuring tools

When we measure things, we are actually comparing an unknown dimension with a known standard. As precision in measuring is of utmost importance, accurate measuring instruments are essential for large-scale production. Without these instruments, accurate and interchangeable parts could not be built and production output would be much lower. Today sophisticated measuring tools are used in all machining operations, from the layout of the different parts right up to the final product inspection.

a Match the following pictures with the names of measuring instruments listed below.

1. feeler gauges
2. folding rule
3. steel square
4. tape measure
5. vernier caliper
6. micrometer
7. dial indicator

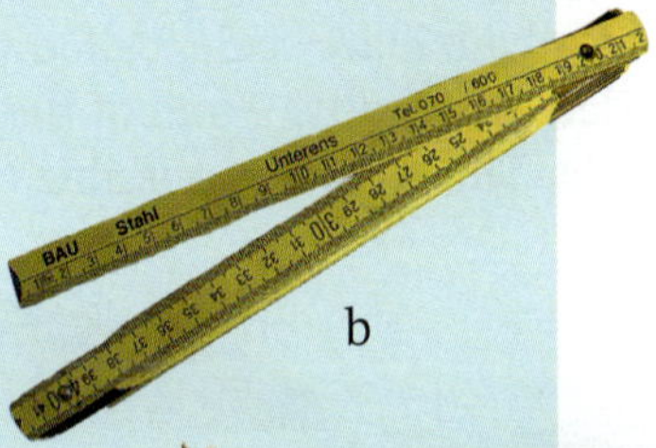
b

c

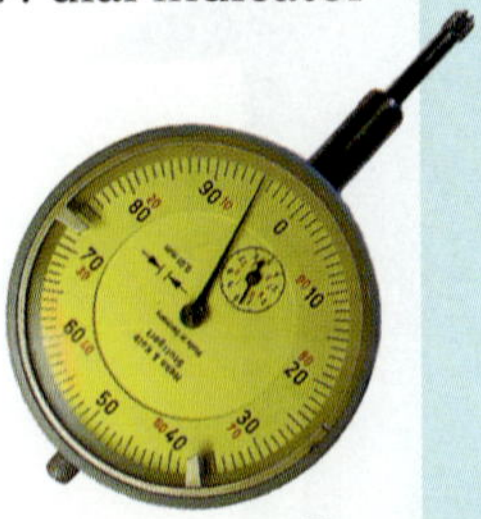
a

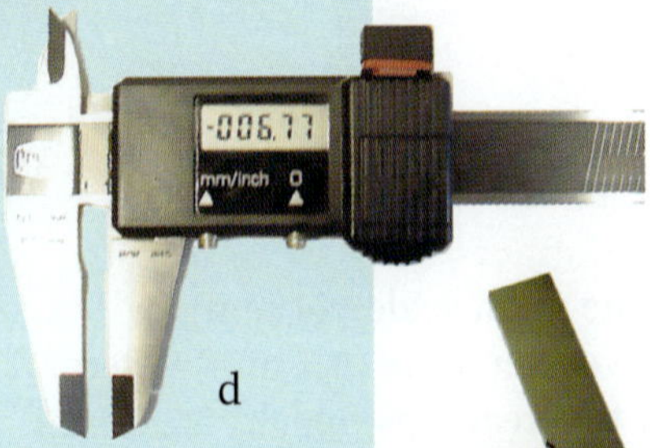

d

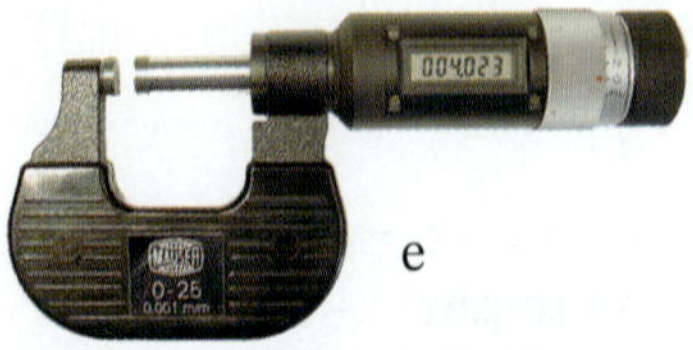

e

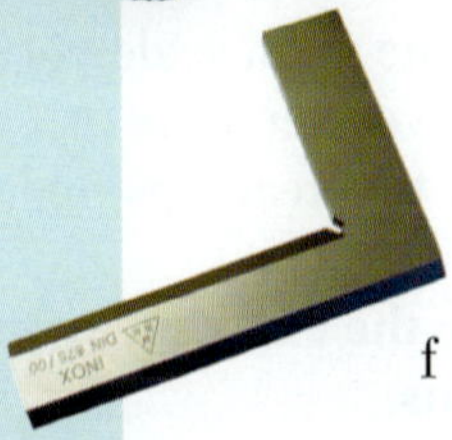
f

g

b Which of the instruments would be used by

1. a joiner to measure the length of a board?
2. a carpenter to measure the length of a wooden beam?
3. a car mechanic to check the distance between the electrodes of a spark plug?
4. a technician to see if a workpiece he has just finished turning has the required diameter?

c Read the following text and match the terms below (1.-8.) with the letters in the illustration of a vernier caliper.

The most versatile measuring instrument is a vernier caliper or sliding caliper. It consists of two pairs of fixed measuring jaws and movable measuring jaws. One pair of jaws is for outside or external measurement: You can measure the width or thickness of a workpiece or the diameter of an axle. The other pair consists of the jaws for inside measurement, or internal measuring jaws. They measure the inside diameter of a ring or a tube. The third part of the instrument is the depth gauge, used for measuring the depths of hollow objects. In order to fasten the jaws you use a clamp screw.

1. fixed jaw
2. movable jaw
3. jaws for inside measurement
4. jaws for external measurement
5. depth gauge
6. main scale
7. vernier or nonius
8. clamp screw.

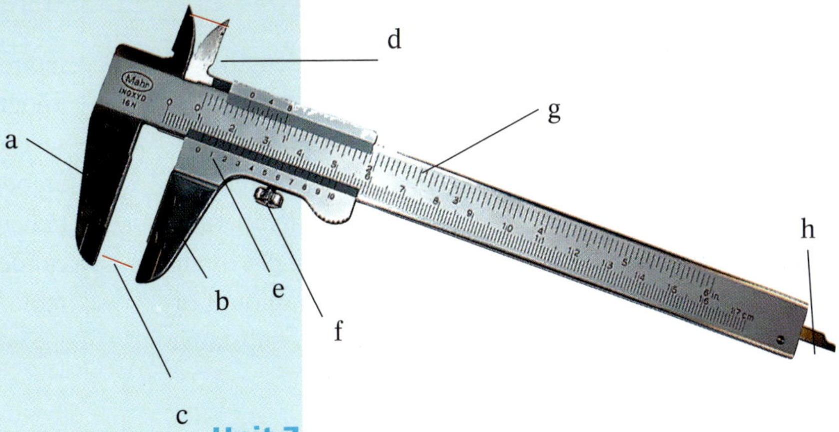

d The procedure for reading off the measure is rather complicated. While adjusting the two jaws, you move two scales against each other. The main scale refers to the fixed jaws and is divided (or calibrated) in millimetres. The small auxiliary scale, called vernier or nonius, refers to the movable jaws and is divided differently: 9 mm of the main scale are represented in 10 parts on the auxiliary scale. This makes it possible to read measurements up to a tenth of a millimetre.

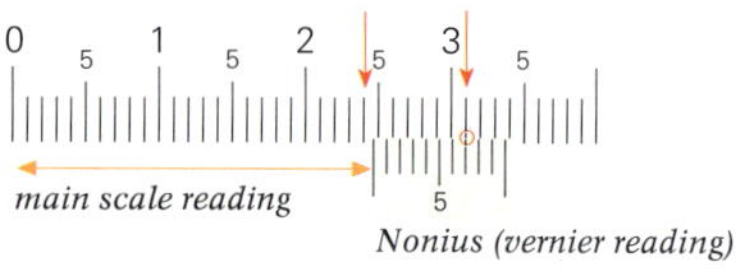

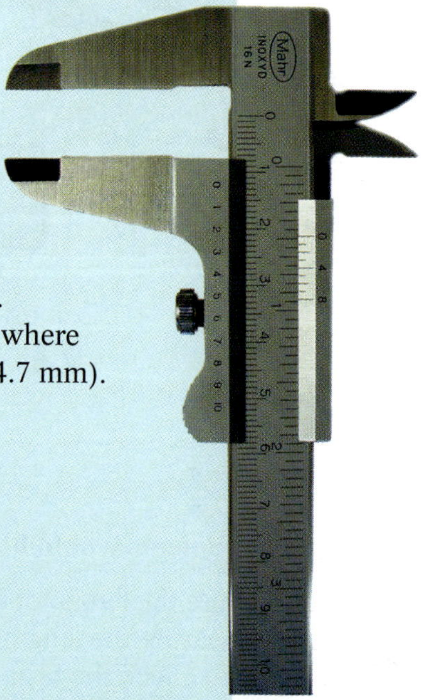

Example:
Read off the whole millimetres on the main scale (24 mm).
Read off the 1/10 mm value on the vernier scale (0.7 mm) where the two scales are aligned. Add the two readings (result: 24.7 mm).

e Study the example above and give the exact measures shown in the following illustrations.

1.

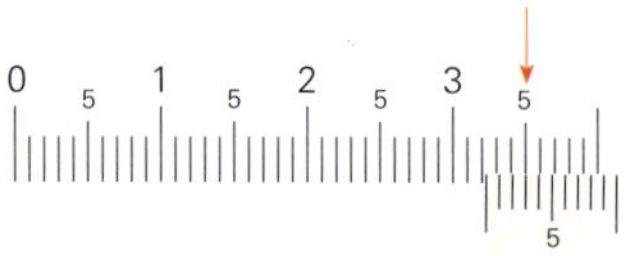

2.

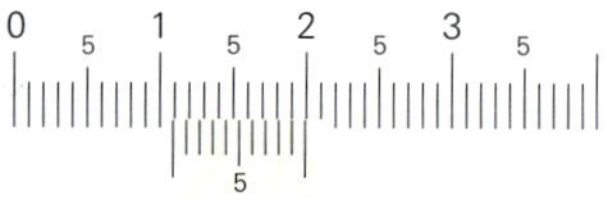

C2 Units of measurement

Here are the basic SI units (SI stands for Système International d'Unités) that have been the basis for measuring all over the world since their introduction in Paris in 1960.

Base quantity	Name	Base unit
length	metre	m
mass	kilogram	kg
time	second	s
current	ampere	A
temperature	kelvin	K
light intensity	candela	cd
amount of substance	mole	mol

- **Look at the table on page 112 and fill in the missing base quantities, names or base units:**

  Metre (m) is the base unit for ①. Electric current is measured in ②. Candela is the name for ③. Mass is measured in ④. K means ⑤ and is the unit for ⑥. The base unit for measuring ⑦ is s. The amount of substance of an electron is measured in ⑧.

C3 ISO standards

The *International Organization for Standardization* (ISO) is a non-governmental organisation which was established in 1947 and which is based in Geneva. The ISO has developed a series of international quality standards, known as the ISO 9000 standards, which are reviewed approximately every 5 years. ISO 9000 was first issued in 1987. The ISO 9000 standards are meant to help companies in all aspects regarding quality procedures for products and services. ISO 9000 registration proves that a company meets the requirements of the standards and that it is able to supply assured quality products or services. ISO-registered companies are recognised around the world for quality.
Businesses applying for an ISO registration will be audited by an authorised agency. The agency checks the conformity of the company with the ISO standards and grants ISO 9000 when the audit has been successfully completed.

In November 2000 a new version of the ISO 9000 family was issued. Some of the major changes were:

- the merging of ISO 9001, ISO 9002 and ISO 9003 into a single standard, ISO 9001
- greater emphasis on continual improvement methods
- greater emphasis on resource planning and management

The new structure of the ISO 9000 standards is as follows:

ISO 9000	Concepts and Terminology
ISO 9001	Requirements for Quality Assurance
ISO 9004	Guidelines for Quality Management of Organizations
ISO 19011	(formerly 10011) Guidelines for Auditing Quality Management Systems

a Give a short summary of the text in French.

b Why is it important for a company to get ISO 9000 registration?

c Give examples of what the ISO has achieved so far. Use any source available. Think of credit and telephone cards, freight containers, paper sizes, etc.)

Example: The ISO has developed international film speed codes.

d Explain in your own words why international standardisation is needed.

C4 *T*otal *Q*uality *M*anagement *(TQM)*

There can hardly be anyone in the manufacturing or service industries who has not yet heard of quality control or quality management. Most industrial companies have implemented quality management systems. Among the first to do so were the car manufacturers and their suppliers. The SMEs – small-to-medium enterprises – were next. Craftsmen, carpenters, car repair shops, plumbers, or the electrician round the corner also need to take this trend into account.

Quality management simply means that the quality, cost and service of a product must meet the customer's needs. A basic principle of quality management is the concept of continuous improvement, with all employees being responsible for the quality of each aspect of their work.

Many schools, technical colleges and private educational institutions now offer special courses to train their students in the idea of quality management and the different tools of measurement, including statistical tools.

a Answer the following questions on the text.

1. Which kinds of companies have installed quality management systems?
2. How does the text define quality management?
3. Where can you learn about quality control?

b Read the following statements on Total Quality Management. Three of the statements are wrong. Which ones?

1. TQM primarily affects the management of a company.
2. A priority of TQM organizations is to meet customer requirements
3. Staff in TQM organizations have less responsibility.
4. Problems are not seen as isolated, but as the result of many causes.
5. Information within TQM companies flows top to bottom and not bottom to top.
6. Employees take part in decision-making processes.
7. Problems are prevented rather than simply being corrected.
8. TQM means that a company is constantly trying to make things better.

c Find more information on TQM. Use any sources available.

D Grammar revision: Gerund

See Grammar Overview in the appendix.

a Put in the correct form of the verb. Note that the verb becomes a noun when –ing is added.

Example:

Car manufacturers are always interested in … about new and lighter materials. (**to learn**).
*Car manufacturers are always interested in **learning** about new and lighter materials.*

1. The manager of our company took great pleasure in … our new plant to the public. (to present)
2. He apologised for … the press waiting. (to keep)
3. He insisted on … them personally on a tour through the buildings. (to take)
4. The politicians said that they were looking forward to … the unemployment figures drop. (to see)
5. They said they were glad to see that the companies were capable of … themselves and … their own new ideas. (to help; to develop).
6. The workers' representative said: "Fifteen months ago we were all in danger of … our jobs; now we are facing a hopeful future." (to lose)

b Put in the correct verb forms.

Note:

remember + gerund	=	se souvenir d'avoir fait quelque chose : I remember repairing the tool.
remember + infinitive	=	ne pas oublier de faire quelque chose : I have to remember to buy a new catalogue.
to stop + gerund	=	arrêter de faire quelque chose : He stopped working.
to stop + infinitive	=	(s')arrêter pour faire quelque chose : He stopped (working) to talk to me.

1. He remembered that he had called the service assistant. He remembered ... the service assistant.
2. She remembers quite clearly that she filled in the form yesterday. She remembers ... the form yesterday.
3. Don't you remember that you gave me the password a week ago? Can't you remember … me the password a week ago?
4. From then on she didn't exchange a single word with him. She simply stopped … (to talk) to him.
5. She looked up from her computer and stopped ... (to type) … (to listen) to what he had to say.
6. She'll never stop … (to love) him.

Unit 8

Energy and the environment

The IEO99 (International Energy Outlook) projects that "every energy source except nuclear power will grow over the 1996 to 2020 forecast period." According to the IEO, renewable energies are not expected to grow as fast as they have in the past, and oil will remain the dominant source of energy worldwide.

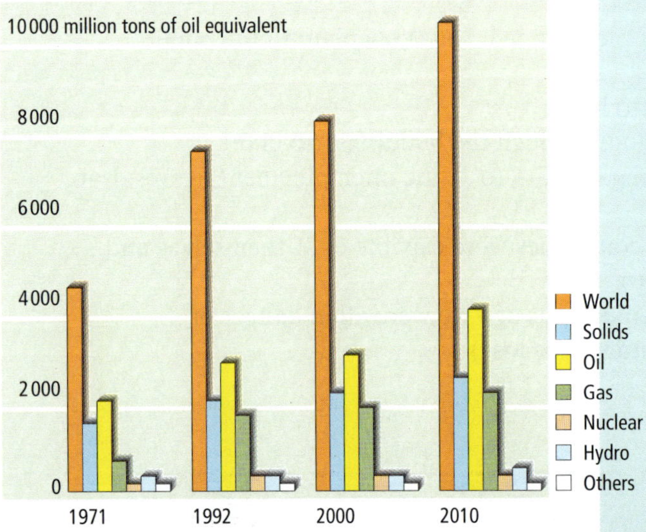

World energy forecast

a **Describe the changes in world energy consumption as shown in the bar chart. Use the phrases below.**

b **Which is the most important source of energy?**

c **Where are the most dramatic changes?**

d **Which source of energy will show only a slight change?**

e **Explain the category 'others'.**

Phrases: Graphs and diagrams

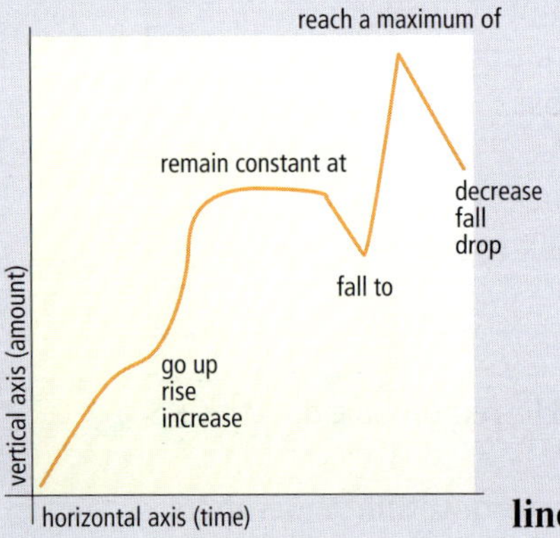

line graph

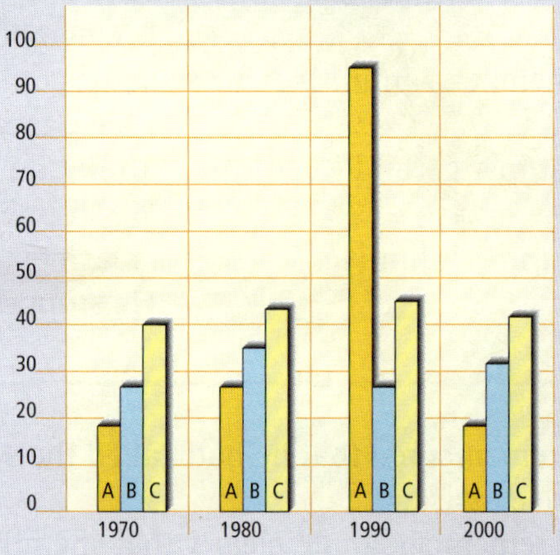

bar chart

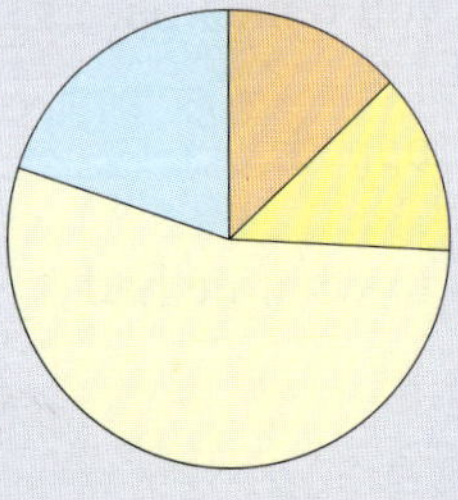

to amount to... %

to take up... %

pie chart

Example:

In 1990 there was a **sharp rise** of sales (A) (orange). (In 1990 sales **rose sharply**).

degree of change

high – enormous(ly)
– dramatic(ally)
– sharp(ly)

– substantial(ly)
– considerable
– considerably

– little
– slightly
low – hardly

A Global warming and the greenhouse effect

The term "global warming" describes the increase in the average temperature on earth. This increase, which is called the "greenhouse effect", can be attributed to a large extent to the rise in carbon dioxide, which is mainly caused by emissions from industry and cars. Coolants in refrigeration and propellants in aerosol cans also contribute with chlorofluorocarbon (CFC).

The greenhouse effect derives its name from the processes that take place in a greenhouse. The glass allows short-wave radiation to enter a greenhouse but prevents outgoing long-wave radiation from leaving it. The result is an increase of the temperature in the greenhouse.

Carbon dioxide, methane and nitrous oxide are natural gases which are essential for life on earth, as they radiate lost heat back to earth. If there weren't this natural greenhouse effect, almost all radiation would be returned to space and the average surface temperature would be around 0° C.

The problem is that with the increasing amount of carbon dioxide in the atmosphere, more heat is trapped and the earth's surface temperature begins to rise.

Some of the effects of global warming could be more violent storms, the spreading of tropical diseases, the melting of glaciers and icecaps (which would result in floods), and increases in pollution levels.

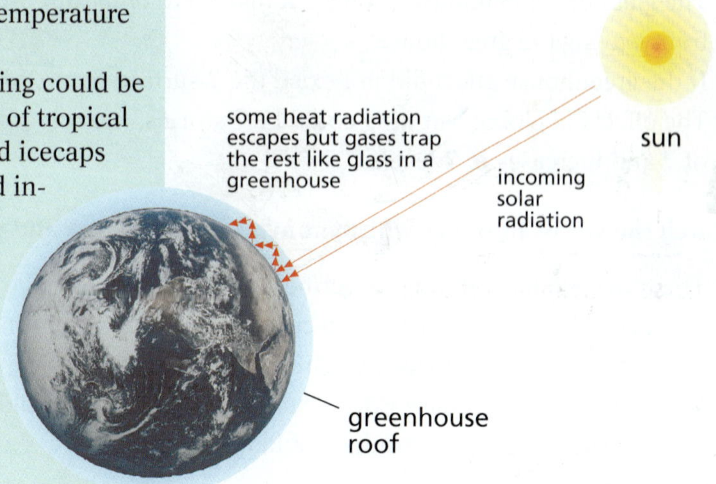

How the greenhouse effect works

Gases that heat up the greenhouse roof:

- carbon dioxide → coal, oil, fire / volcanic eruption
- CFC gases → coolants, aerosols
- methane + nitrous gases → cattle, fertilizers

a Explain the terms "Global Warming" and "Greenhouse Effect".

b Answer the questions on the text.

1. Suppose there were no greenhouse effect – what would be the average temperature on earth?
2. Which atmospheric gases are responsible for the greenhouse effect?
3. What are the main causes for these atmospheric gases?
4. What happens when the sun's short-wave radiation hits the earth?
5. Which atmospheric gases are caused by man?
6. Which atmospheric gases are caused by nature?
7. What might be the effects of global warming?

c Complete the following sentences with words from the text.

1. The term "global warming" refers to an increase in the ? temperature on earth.
2. Scientists have observed an increase in the earth's ? temperature.
3. This increase is called the ?.
4. The glass of a greenhouse allows ? radiation to enter but prohibits outgoing ? radiation from leaving the greenhouse.
5. If the greenhouse effect did not exist, the ? surface temperature would be around 0°C.
6. The effects of global warming could be ? storms, the spreading of ? diseases, the melting of ? and increases in ? levels.

d Match the words in italics with synonyms from the text and write them down.

1. The term "global warming" describes the *rise* in the average temperature on earth.
2. The rise in temperature is mainly caused by *pollutants* from industry and cars.
3. This effect is *referred* to as the "Greenhouse Effect".
4. Without the "Greenhouse Effect" almost all radiation would be *reflected* into space.
5. Some *results* of global warming could be disastrous.

e Work in pairs and discuss what people can do to reduce the "Greenhouse Effect".

Example:
People should use more public transportation instead of cars.

A1 Change of temperature

Temperature anomaly in the 20th century:

Change of annual global surface mean temperature in relation to carbon dioxide concentration:

year	1900	1910	1920	1930	1940	1950	1960	1970	1980	1990	2000
Change of temperature (°C)	- 0.18	- 0.3	- 0.2	- 0.1	0.08	- 0.05	0.08	0.0	0.25	0.4	0.7
Carbon dioxide concentration (ppm*)	290		300		315		320		330		360

* ppm = parts per million

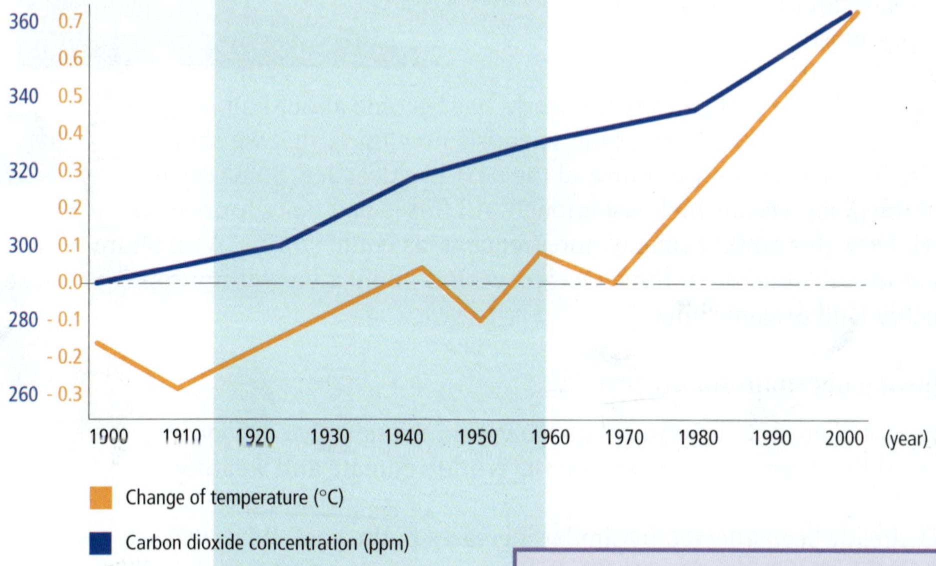

a **Describe the changes in the global mean temperature during the 20th century (orange line).**

b **Describe the carbon dioxide concentration in the 20th century (blue line).**

c **Explain the relation between the global mean temperature and the concentration of carbon dioxide.**

Use the phrases on the right and on page 116 to describe the graph.

d **You are a member of an environmental interest group which discusses possible reasons for this trend. Find arguments for a discussion at a meeting.**

Phrases: Facts and opinions

Expressing facts	It is quite obvious … There is no doubt that … It is evident that … On the one hand, … on the other hand, … The main reasons for … are … There are several factors which contribute to … First of all we have to consider … Second(ly) we should not forget …
Giving reasons	This is as a result of … This is due to … (the fact that)
Describing the personal view	From my point of view, … In my opinion, …
Summing up	To sum up, … Summing up all the arguments, … it can be said that …

A2 Rise in sea levels

In the next 50 years the mean temperature of the earth could rise by between 1° and 4° C. A rise of 4° C is about the difference between the global temperature now and the temperature of the last ice age. This increase in temperature would first of all alter the world's climate and weather patterns. The tropics would get a little warmer and the polar regions could become dramati-cally warmer. In the past 20,000 years life on earth was able to adjust to this change. In future we will not have another 20,000 years to adjust. The most pessimistic calculations predict that we have less than 50 years to do so. Water expands with heat. In the past century the world has become about half a degree warmer. Sea levels have risen by 10-15 cm. Many scientists now think that we should prepare for a 1-m rise in sea levels in the course of the next five decades. Some say it could be 2 metres if the polar icecaps melt fast enough. All this is bad news for people who live at sea level. Droughts could become more frequent, as could storms. Agriculture could be disastrously hit; so could the forests. Ocean currents could alter, with Europe becoming colder as the Gulf Stream shifts.

a Answer the following questions on the text.

1. How could the global mean temperature change within the next five decades?
2. What effects would such an increase have on the world's climate and weather patterns?
3. The world has already been affected by similar increases in the past. What will be the difference this time?
4. What are the possible effects of an increase of the average temperature by half a percent?
5. What are the most pessimistic calculations for the next 50 years?
6. In which way could agriculture be hit?
7. Why could Europe become colder?

b Match the words in italics with synonyms from the text and write them down.

1. In the next 50 years the *average* temperature of the earth could rise.
2. An *increase* of up to 4° C is *forecast* by the most pessimistic calculations.
3. Such an increase in temperature would *change* the climate on earth.
4. In the past 20,000 years life on earth has been able to *adapt* to this change.
5. Many scientists *are* now *of the opinion* that a 1-m rise in sea levels could become reality.
6. *Long dry periods* could become more frequent.
7. Agriculture could be hit *catastrophically*.
8. Europe could become colder as the Gulf Stream *changes its direction*.

c Find the corresponding nouns, verbs, adjectives, adverbs in the text.

1. different	(noun)		5. prediction	(verb)
2. climatic	(noun)		6. preparation	(verb)
3. drama	(adverb)		7. frequency	(adjective)
4. pessimism	(adjective)		8. disaster	(adverb)

d Translate the text into French.

B Renewable sources of energy

a Complete the following text using words from the box.

renewable – derived – depleted – converted – combustion – efficiency

① energy is ② from resources that are generally not ③ by humans, such as the sun, wind, and water movement. These sources of natural energy can be ④ into heat, electricity and mechanical energy in several ways. Some technologies for conversion of renewable energy such as hydropower, biomass, and waste ⑤ have already been applied for many years. Other conversion technologies, such as wind turbines and photovoltaics, are already well-developed, but have not achieved technological ⑥ so far. Although geothermal energy is produced from geological sources, it is generally regarded as a renewable energy resource. Commercial nuclear power is not considered to be a renewable energy resource.

b Work in pairs. Describe the pros and cons of three renewable sources of energy.

c Discuss in groups: Why does nuclear power not belong to the renewable sources of energy?

B1 Solar power

Solar power is an environmentally friendly technology. The sunlight is converted into electricity without any polluting side effects. We can use the sun in two ways to produce energy: solar cells and solar panels. Photovoltaics (PVs), generally called solar cells, are semiconductor devices that convert light directly into electricity. They are made of thin slices of silicon. Two different types of silicon layers are connected by a wire. When light shines on them, an electric current is produced. A single solar cell produces only very little current, approximately 0.5 volts,

but when many individual cells are connected in series, as in a battery, they generate a useful amount of power.

Solar cell technology was first used in space, but is now increasingly used in other applications. Today pocket calculators, watches and even cars and communications satellites are powered by solar energy.

Solar panels, on the other hand, collect the heat of the sun to supply us with hot water. Pipes under a glass are fixed to a black plate. The sun heats a liquid in the pipes and this heats up a tank of water. Huge solar furnaces can use the sun's heat to generate electricity. Curved mirrors collect the sun's heat and concentrate it onto a furnace, where the heated water is converted into steam. This drives a turbine producing electricity like an ordinary power station.

Solar technology has improved steadily over recent years and, as a result, the cost of solar energy has decreased considerably.

a Answer the following questions on the text.

1. Which solar power technologies can be used to produce energy?
2. How can solar panel technology be used to generate electricity?

b Describe how energy is generated in solar cells.

c Explain solar panel technology.

d Read the advertisement on page 123 and find the English translations for the following French words.

1. panneaux solaires
2. garantie
3. matériel
4. de pointe
5. onduleurs
6. adaptés à
7. disponible
8. chargeurs de piles

e You work for a French company specialising in solar technology and have just studied the advertisement from the Internet on the right.

Write a letter to *Solar Products Ltd.* enquiring about :

- range of products (solar cells, solar panels, etc.)
- technical details
- costs/efficiency
- terms of delivery and payment
- brochures and price lists
- discounts

See letter layout and phrases in Unit 2.

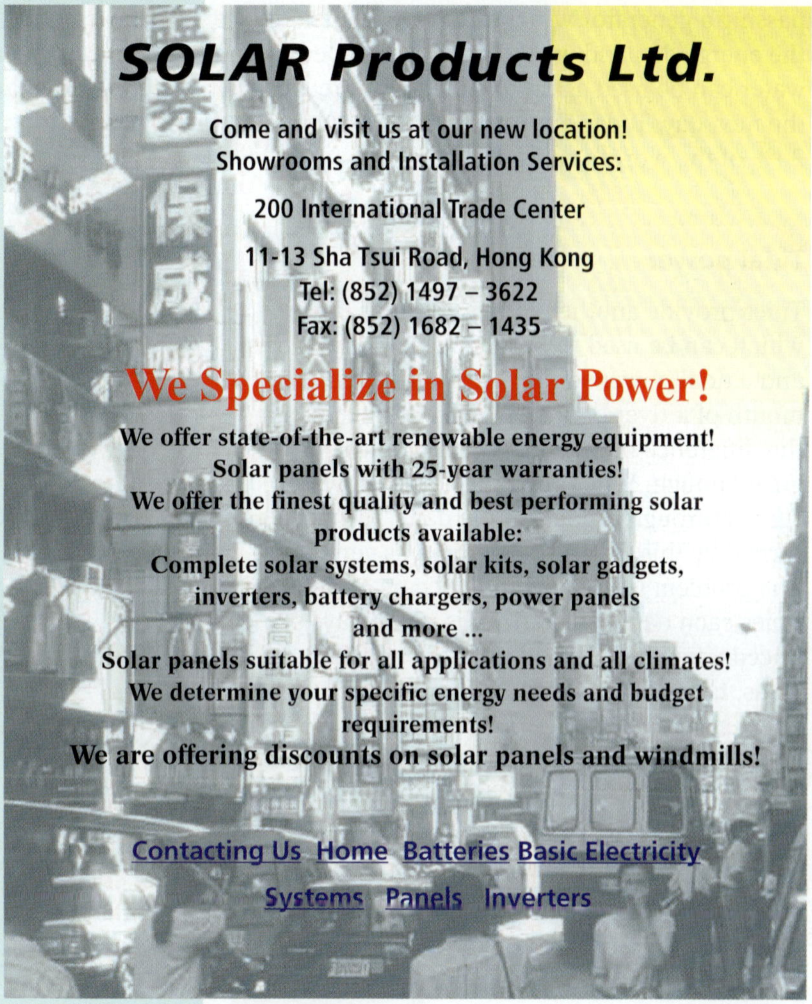

B2 Water power

Hydroelectric power

There are different ways to generate electricity with the help of water power. Water which is stored behind a dam possesses potential energy. This potential energy can be changed into other forms of energy. In a pumped storage system, pipes run between two reservoirs at different levels. When water is allowed to run down from the higher level to the lower, it passes through a turbine and generates electricity. When there is a high demand for energy during peak hours, water flows down from the upper reservoir to the lower reservoir,

passing a generator which produces the energy we need. During the night, when the energy demand is low, water is pumped up to the upper reservoir. Then the water can be used again to generate more electricity when the demand is higher the next day. Today approximately 430 hydroelectric power stations produce almost 20% of the world's energy demand.

Tidal power

Tides provide another source of water power which can be used to produce electricity. A dam and a tidal power station are built across the mouth of a river or an estuary in a place where the difference between low and high tide is great enough. When the tide comes in, water rushes through tunnels in the dam. Turbines are turned by this flow and electricity is generated. The problem is that high tide occurs at different times each day, so electricity is not always produced when it is needed most. Moreover, tidal areas, home to fish, birds and other animals, are affected by the dams.

Wave power

Scientists are now exploring ways of using the rising and falling motion of waves in the open sea. This is a reliable and cheap source of power. In order to generate electricity a certain area of the sea is covered with floats which are connected with moveable parts. The kinetic energy of the up-and-down movements of the floats is transformed into electricity as in a dynamo. The disadvantage of wave power is that the amount of energy produced is very low. So large parts of the sea would have to be covered in order to generate enough power.

a There are three principal ways to generate electricity with the help of water.
What are these three technologies?

b Describe the basic technology of a pumped storage system.
Why do environmentalists often dislike this technology?

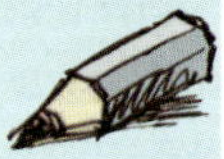

c Explain the most significant disadvantage of tidal power.
How is electricity produced with the help of tidal power?

d The wave motion can also be used to generate electricity.
What is the disadvantage of wave power?

e Find synonyms for the following words in the texts on the water power.

1. various
2. produce
3. furthermore
4. require
5. need
6. kept in reserve
7. nearly
8. safe, dependable
9. drawback
10. movement

f Give a summary of the texts in French.

B3 Wind power

Wind has been harnessed for centuries. Ships have sailed around the world on wind power alone. Windmills have used the power of the wind for grinding corn and pumping water. Today we use wind energy to generate electricity with so-called aerogenerators. Aerogenerators are huge blades (up to 20 m long) which are mounted at the top of slim towers. The blades spin in the wind and turn a generator which produces electricity. Aerogenerators are placed on exposed, windy sites, often in large groups called wind farms. They do not pollute the atmosphere, but the force of the wind is unpredictable and only a few regions of the world are suitable for this kind of renewable energy.

a Wind has been used for centuries as a source of energy. Give examples.

b Which countries and which landscapes are ideal for the use of wind power?

c What are the disadvantages of wind power?

d You are in different project teams planning to open up hotels on an island in the Pacific. The government insists on an environmentally friendly way of building. There are forests in the southeast, a hot and sunny climate and areas with strong winds from the west.

Work in groups. Discuss the following questions in each group and list your arguments and answers on the blackboard. Then compare the results.

- What material would you use to build the hotels?
- Where would you build your hotels on the island?
- Which sources of energy could you use to generate electricity?
- Do you see any environmental problems that may arise?
- How could the island be affected by your plans?

e **What are the current trends in renewable energies? Work in groups and check any sources available. Present your findings to the class.**

C Nuclear fission and nuclear fusion

C1 Nuclear fission

Atoms are made up of different particles which contain a huge amount of energy. Strong forces exist between the particles of the nucleus of an atom. These forces can be harnessed to generate energy – and this is the principle that has been applied in nuclear power stations.

The problem with nuclear power stations is that they create radioactivity and huge amounts of nuclear waste, which can become extremely dangerous for the environment and all life on earth in the case of accidents (e.g. Chernobyl), earthquakes etc. The nuclear waste problem is far from being solved, which is not surprising, given the fact that plutonium in the waste is said to remain radioactive for up to 240,000 years – and radioactivity is known to cause cancer.

C2 Nuclear fusion

Nuclear fusion is different. All stars including the sun get their energy from nuclear fusion. In nuclear fusion two or more nuclei fuse together and set free a huge amount of energy. In the sun, for example, hydrogen nuclei are continually fusing to form helium nuclei. In this process some mass is lost and converted to an enormous amount of energy. It is said that, in contrast to nuclear fission, radioactive wastes created by nuclear fusion do not contain any long-living isotopes and that no gases are produced that are detrimental to the environment.

So far, however, nuclear fusion has not been a practical or economical way to generate energy on earth, and many scientists think it is more likely that other energy sources will replace fossil fuels in the near future before nuclear fusion becomes a real alternative.

INFO-BOX: Tokamak

In order to start a reaction which leads to a fusion of nuclei, a temperature of millions of degrees is necessary. Most fusion research uses a machine called a tokamak. This is a doughnut-shaped vessel containing the gas to be fused, which is called plasma. No material can cope with such heat, so magnetic fields are used to keep the hot plasma away from the vessel walls.

A powerful electric current flows through the plasma consisting of deuterium atoms. This heats the plasma and produces a magnetic field which compresses the plasma and keeps it away from the walls of the vessel. The high temperature and pressure of the plasma causes fusion. So far, researchers have been able to heat the gas to millions of degrees for only a few billionths of a second – not long enough to start a fusion reaction. But some scientists are sure that they will succeed one day – thus producing energy in abundance.

a Some facts about nuclear fusion: Put the sentences into the correct order.

1. fusion accelerator / most powerful / the / in Albuquerque / is
2. is necessary / to start / a 100-trillion-watt pulse/ a reaction / of electricity
3. towards / deuterium gas / the pulse / of / a pea-sized pellet/ is directed
4. millions of degrees / the plasma / for a few / is heated to / billionths of a second
5. not / so far / succeeded / all efforts / a controlled nuclear fusion / have / to produce

b Nuclear fusion might one day solve some of our energy problems, but there are still enormous problems to be solved. Name some of the problems.

c Why are nuclear fission and nuclear energy no longer seen as a way to solve our energy problems? Think of additional arguments which are not mentioned in the text.

C3 Project work. Internet search: Future sources of energy

a Form groups of 3 to 4 students and discuss in class which possible future sources of energy might be of interest.

b List the sources on the board and decide which source you want to talk about.

c Search the Internet for information about the source of your choice.

d Summarise the information for a presentation in class.

e Present your results to the class.
Follow the basic principles of presentation on the next page:

INFO-BOX: Basic principles of presentations

Checklist Presentation

Audience
- Address the audience directly.
- Involve the audience.
- Speak from headlines rather than reading a text.
- Use a pointer to help the audience follow your presentation.

Structure
- Start with an overview of the areas you are going to cover.
- Break the main body of the presentation into several different sections.
- Finish with a summary of your main points.
- KISS (Keep it short and simple!).

Equipment and Media
- Make sure the equipment is working and that you know how to operate it.
- Vary the media to maintain the audience's interest and attention.
- When using presentation software, don't use too much animation.
- Don't stand in front of the board / screen.

Voice and Language
- Speak clearly.
- Check that everyone can hear you.
- Use language that is appropriate to your audience.
- Don't use too much jargon.

Body language
- Have regular eye-contact with all the audience.
- Don't move around too much.
- Don't talk with your back to the audience.
- Avoid repetitive gestures.

D Grammar revision: Participle (present and past)

See Grammar Overview in the appendix.

a Supply the correct form of the verbs in brackets.

Solar panels transform sunlight into electricity, (1. to guarantee) a high energy output. (2. to be) very resistant to the effects of weather, they are quite durable. They have become much cheaper and are included in more of the new homes (3. to build) today. Sunlight (4. to convert) into heat is not as effective. (5. improve) the insulation of your home is another way to reduce energy costs. Home owners (6. to plan) to install solar technology should balance costs and efficiency. In any case, before (7. to renovate) your home, you should see an energy consultant.

b Rewrite/combine the following sentences using participle constructions (-ing or -ed).

1. *Home owners who plan to install solar panels are advised to compare several offers.*
2. Scientists have studied the greenhouse effect. They attribute it to the rise in emissions from industry and cars.
3. After they had gathered information from the Internet, they started work on their presentation.
4. Before the secretary ordered the battery charger, she asked for a price list.
5. We have not achieved a major breakthrough in fusion. Therefore, we must continue with research in renewable sources of energy.
6. Scientists who work in nuclear power stations are often supporters of this technology.
7. The wind farms which have been built along the coast generate enough energy for 5000 households.
8. Water that is stored behind a dam possesses potential energy.
9. We know that pollution is not only caused by cars. We should take steps to reduce the emissions.
10. When you describe renewable energies, you must not forget water power.

Example:
1. Home owners planning to install solar panels are advised to compare several offers.

c Translate into English using participle constructions.

1. Avant de commencer l'installation du panneau solaire, le propriétaire de la maison a consulté la notice.
2. Tout en gagnant en popularité, l'énergie solaire est devenue une source d'énergie sans impact sur l'environnement.
3. Ne comprenant pas l'anglais, il n'a pas pu lire le mode d'emploi du panneau solaire.
4. Nous avons commandé le panneau solaire suite à la lecture de l'annonce.
5. Il écoutait de la musique tout en lisant l'annonce.

Unit 9

Business trips

a Name the documents, forms, etc. above and say what they are used for.

b What do you have to do when you plan a trip abroad?
Make a list of possible travel arrangements.

A A Business trip

A1 Planning a trip

PHILADELPHIA – THE NEW PHILACON VALLEY

Who talks about Silicon Valley? We don't!

People like to call us Philacon Valley – Technophilly – Silicon Valley Forge or simply @Philly. After a powerful revolution more than two hundred years ago our city is ready for the next one: the revolution of the mind.

Remember – it was this city that made the computer age possible: We made ENIAC in 1946 (or our geniuses at the University of Pennsylvania did).

a Look at the advertisement above. What does it promote?

b Explain the source of the words Philacon Valley, @Philly, Technophilly.

c KeySys is a French company that produces membrane switches and membrane keyboards, which are needed to input data into machines or computers or to operate computers or CNC machines in a damp or dirty environment.
Read the following text and find the correct forms of the verbs in brackets.

As KeySys (① want) to gain a foothold in the overseas market they (② decide) to open a US branch. With the help of American advisors, Smith Consulting Inc., they (③ agree) on Philadelphia as the place for their future branch during a meeting last year. The plans for the future plants (④ already finalize), a large enough property in a nice suburban area near Trevose (⑤ buy), and all the other paperwork has been successfully completed. KeySys expects that the new plant (⑥ finish) by next spring. With the help of the Greater Philadelphia Chamber of Commerce they (⑦ arrange) to present themselves at the Tech Expo USA, a job fair at the Philadelphia Convention Center, where they (⑧ hope) to find new employees.

The British nationals Valerie Piskei from personnel and Chris Fehily from the technical department (⑨ choose) only a week ago to represent their company at the Tech Expo USA in Philadelphia.

d Listen to the following dialogue and take notes of important points.

e Listen to the dialogue again and say which of the following statements is not given or is wrong.

1. Chris prefers a stopover in New York or Boston.
2. Valerie would like to book a hotel close to the airport.
3. Valerie and Chris agree on renting car.
4. They don't need an international driver's license.
5. Valerie would like to go to an open-air concert.
6. Chris is fond of sea food.

f Make a list of all the things Valerie, Chris and their secretary have to do and to check before and during their business trip to Philadelphia. Use the verbs have to / must / should('nt) / need(n't).

check passports

plane tickets

taxi to the airport

cultural events in the evenings

Example:

First they need to call a travel agency to book a suitable flight. Then they have to arrange for ...

A2 Making hotel reservations

a Valerie and Chris left their secretary, Laura Martin, a memo, asking her to find a suitable hotel that offered the facilities listed in the memo.
Read the memo and compare the two hotel advertisements on page 133.
Which hotel is more suitable for Valerie and Chris? Give reasons.

MEMO De : Valerie Piskei A : Laura Martin Date : 5. 6. 2004

Objet : Hôtel à Philadelphie

Merci de chercher un hôtel à Philadelphie pour Chris Fehily et moi-même (pour un voyage à la Foire fin août). L'hôtel doit offrir les services suivants : machine à café dans la chambre, fer à repasser/possibilité de repasser, séchoir à cheveux, connexion Internet, salle de gymnastique, literie hypoallergénique, chambre non-fumeur, laverie, télévision avec câble et satellite, location de voiture, journaux du jour, taxi pour l'aéroport.
Merci
Valerie Piskei

DENTON INN PHILADELPHIA/ AIRPORT, PA.

4200 Bartram Ave
Philadelphia, PA 19153 US
Ph: 215-966-2983
reservations@dentoninn.com
1-800-DENTON (426-7866)

Area Information:

The hotel is located one half mile from the Philadelphia Airport, with easy access to I-95 North and South. We are seven miles from Downtown, Independence Mall, Convention Center and the Historic District. We are only minutes from the city's many major attractions with ideal access to all major highways.

Philadelphia Intl Airport	1/2M N,	Shuttle (No charge).
Philadelphia Downtown	7M N,	Direct rail shuttle.
Convention Center	7M N,	Rail shuttle.
Historic District	7M N,	Rail shuttle.

Hotel Features:

Facilities available at or near hotel
*Indicates extra charge

- Courtesy airport shuttle service is available
- Coffee maker in Room
- Connecting Rooms
- Cribs available
- Exercise gym
- *Audio-Visual Equipment
- Hairdryers available
- *Valet services
- Iron/ironing boards in every room
- Free breakfast bar
- Laundry service
- Car rental office
- Free local phone calls
- Dataports/Internet in each guest room
- *Meeting/banquet facilities
- Pool
- Satellite television

Hotel Information:

- Check-In: 15:00
- Check-Out: 12:00
- Pool: Yes
- Pets: No
- Credit Cards Accepted

All-Seasons Hotel

Logan Square
Philadelphia, Pennsylvania
19103-6933, U.S.A.
Telephone: (215) 966-6948
Guest Fax: (215) 966-6970
info@allseasons.com
Philadelphia International Airport: 25 min.
Downtown 5 min.

Hotel features:

Laundry/dry cleaning (overnight)
Newsstand and sundries
No-smoking rooms
Parking
Pressing (one hour)
Shoe shine (overnight), complimentary
Valet parking
Hairdryer
In-room exercise equipment available
In-room safe
Non-allergenic foam pillows available
Private stocked bar/refrigerator
Coffee maker in each room
Health/fitness facilities
Jacuzzi/whirlpool, massage available
Pool, sauna/steam
24-hour business services (computer, Internet connection, voice mail, fax, photocopying, delivery)
Airline ticketing service/Car rental desk/ Airport shuttle service
Audio-visual equipment
Meeting/function space
Translation services
All-news cable TV, clock radio
Check-in: any time
Check-out: 12:00

b What features would you look for if you were booking a hotel? Answer in full sentences using the following phrases and prompts.

The hotel should have / shouldn't have / should be near / shouldn't be ...

The hotel must have / needn't have ...
A ... (e.g. hairdryer) should be available.
It would be great to have / it would be fun to have / it would be very convenient to have ... (e.g. laundry / fitness facility)
Ideal for me would be a hotel with / to be in a hotel with ...

- close to the airport
- limousine / shuttle service to the airport / to city centre
- close to the centre of the city
- in a quiet area
- a bathroom / a shower / pool / exercise gym room
- cable-TV
- refrigerator / mini bar / coffee maker in the room / hairdryer / iron /telephone / dataports
- conference room
- car rental desk
- valet services / valet parking / laundry service

c Work in groups. Find information about hotels in different countries (e.g. from the Internet). Then compare the hotels as shown above. Which hotel(s) would you choose for a business trip? Which hotels would you choose for a family holiday? Give reasons.

d Valerie and Chris have decided on the All-Seasons in the centre of the city. They have left their secretary the following note asking her to send a fax to the hotel. The details are in the memo:

MÉMO De : Chris Fehily A : Laura Martin Date : 8. 6. 2004

Merci de réserver les chambres suivantes à l'hôtel All-Seasons de Philadelphie :
2 chambres calmes avec lit simple et douche.
Pour les détails, voir le mémo du 5 juin (page 132).
Arrivée : 25 août à 16h30.
Départ 31 août à 9h00.

Merci !
Chris

Write a fax or an e-mail that contains the data given above. Use the phrases on the next page.

Phrases: Booking a hotel room

We would like to book a single (two single) room(s) / a double room for … nights from … to …
… with shower / with (en suite) bathroom / nonsmoker / quiet room overlooking the sea / park …).
What are the hotel rates?
Is breakfast / service … / Are meals included in the price?
The day of arrival is … / We will arrive on …
The day of departure is … / We will be leaving on … (day) at (hour).
Is there a limousine / shuttle service to and from the airport?
Could you confirm our booking as soon as possible?
Do you accept … credit cards?

e The next morning Chris finds the following fax on his desk from the All-Seasons Hotel. Note down the most important facts in French.

Dear Ms Martin,

Thank you very much for your fax of … . We are sorry to tell you that the rooms you wanted to book will not be available as August 29 is during Labor Day Weekend, and we are totally booked up.
However, we have made arrangements with the Royal Excelsior Hotel, 6600 Bartram Ave, Philadelphia, which is also closer to the airport. They have two single rooms available during the time in question.
If our suggestion suits you, please contact the hotel. May we advise you not to lose too much time, given the circumstances.
Thank you for doing business with us.

Truly yours,
Carlos Montinez
Booking Office

INFO-BOX: National Holidays USA

Here is a list of U.S. holidays: religious and commemorative.
During the 1970s, Congress established five Monday holidays. The schedule gives American workers five extended weekends at different times of the year:

Martin Luther King Day – third Monday in January
President's Day – third Monday in February (takes the place of Lincoln's birthday – Feb 12, and Washington's birthday – Feb. 22)
Memorial Day – last Monday in May (used to be May 30)
Independence Day – July 4th
Labor Day – first Monday in September
Columbus Day – second Monday in October
Veterans Day – November 11th
Thanksgiving – fourth Thursday in November
Religious holidays are only **Christmas Day** and **New Year's Day**.

f **Compare the list above with the German national holidays.**

g In addition to the legal public holidays, the average American employee has ten days of paid vacation a year, making just two weeks (including the weekends) to actually take a vacation trip. Many employers grant a third week after a certain number of years, and even a fourth some years later. **How does this compare with the number of vacation days allowed in France?**

h After reading the fax, Chris calls the Royal Excelsior Hotel right away. **Put the following dialogue in the right order. Then take the roles of the receptionist (Student A) of the hotel and Chris (Student B) and play the dialogue.**

Receptionist

- I am sorry, but this is the Royal Excelsior Hotel, would you like me to give you the number of the All-Seasons Hotel?
- Royal Excelsior Hotel, Mike Turner speaking. Can I help you?
- Yes, I've got that. Please hold the line, I'll be right back. … Yes, we had a call from the All-Seasons Hotel. The dates are the 25th as the day of arrival and the 31st of August as the day of departure. Two single rooms.
- Oh, I am sorry to hear that, when did you book for?
- Oh, now I understand. I really have to apologise. I'll check with the booking department right away. Who did you say made the arrangement?
- I am afraid not, one is with bath.
- No problem, sir; there is always a limousine parked at the west exit, and if there shouldn't be one use our phone right next to the booth.
- You're welcome, sir. And if you should have any more questions concerning your stay or the services of our hotel make sure to call us right away.
- Bye, and have a nice trip, sir.

- Yes, my name is Chris Fehily, I am calling from France. We had some problems with a reservation …
- Well I am certainly glad we have that settled! Thank you very much.
- No, I am sorry we didn't book at your hotel, we wanted to book at the All-Seasons Hotel for August 25.
- There seems to be a misunderstanding; let me explain. We requested two single rooms at the All-Seasons Hotel. This morning they sent me a fax telling me that they are booked up due to the national holiday. They said they had made arrangements with you for August 25 to August 31, and they told me I should call you right away.
- On my fax here it says Carlos Montinez.
- Are they single rooms with shower?
- That's okay. – And can you pick us up from the airport?
- We'll do that. Thank you and good-bye.

A3 Booking a flight

a **Look at the following schedule for flights to Philadelphia.**

Flight number	weekdays / weekends	departure from	local time	stopover	arrival in	local time
AC 651	Saturdays only	Paris	11.45	Toronto	Philadelphia	18.09
AF 883	daily	Paris	15.20	direct	Philadelphia	17.40
BA 424	daily	Paris	14.40	Heath Row	Philadelphia	?
US 352	daily	Paris	13.20	direct	Philadelphia	?
AF 697	daily	Paris	17.10	direct	Philadelphia	?

b **Valerie and Chris would like to arrive at their hotel in Philadelphia at around 4.30 p.m. Which flight should they choose to be on time?**
Give the approximate arrival times in Philadelphia for the last three flights above.
(If necessary, check the time zones in any source available.)

c **If you have access to the Internet, find out about flights from your nearest airport to different places in the world. Then let the class guess the flying times and time differences before giving the information.**

d **Role play. Create a short dialogue between a customer (Student A) and a ticket agent (Student B). Use the information and the phrases below.**

Ticket agent:

Customer:

has only a seat on an early morning flight leaving at 7.30 a.m

stopover in New York

no window seat, just an aisle seat / row 32

express check-in not possible for international flights

wants to fly to Boston on July 15th / doesn't want to leave too early

prefers a non-stop flight

would like to have a window seat if possible in one of the front rows

asks if he/she could go through an express check-in to avoid standing in line

Phrases: Booking a flight

Possible questions / answers (passengers):	*Possible answers / questions (ticket agent):*
• I'd like to know what flights are available to ... for ...	• The first flight possible is / the only flight is from ... at ... o'clock.
• I'd like to fly first class / business class / economy class.	
• Is there a later plane?	• No, I'm sorry, the next flight is fully booked up.
• Is it a direct flight?	
• Are there any stopovers?	• Flight no. 455 has a stopover in ...
• I would like a window seat.	
• No, I don't want a window seat, I'd rather have an aisle seat.	• What kind of seats would you like?
• Could I have a seat in one of the front rows, please?	• Front or back rows, madam / sir?
• Do you have express check-in for passengers with only hand luggage?	• Express check-in is possible only for domestic flights.

e Work in pairs. Create further dialogues using the phrases above. Make up any missing details.

B Arriving in Philadelphia

B1 At the airport

Shortly before landing at Philadelphia International Airport the flight attendants hand out forms to the passengers that they have to fill in and hand to the immigration officers.

a Complete the excerpt from an immigration card by matching the information below with the numbers in the form.

- KARIN
- 06.05.72
- LH 2479
- 3346899920 B
- 50E. BROADWAY
- KRÜGER
- GERMANY
- FRANKFURT

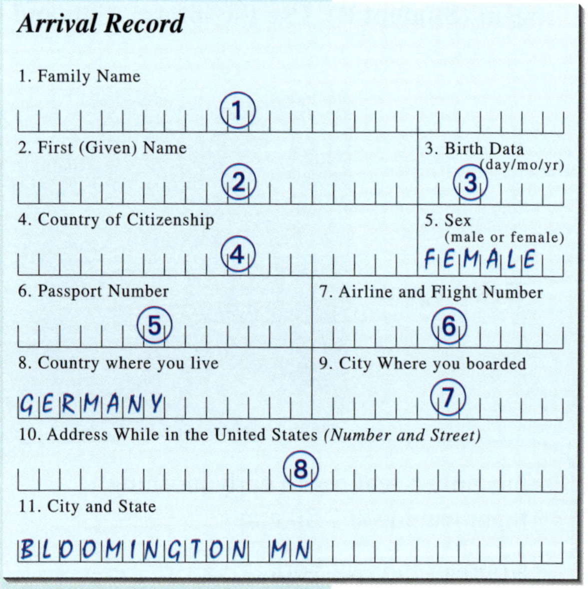

Arrival Record

1. Family Name (1)
2. First (Given) Name (2)
3. Birth Data (day/mo/yr) (3)
4. Country of Citizenship (4)
5. Sex (male or female) FEMALE
6. Passport Number (5)
7. Airline and Flight Number (6)
8. Country where you live GERMANY
9. City Where you boarded (7)
10. Address While in the United States *(Number and Street)* (8)
11. City and State BLOOMINGTON MN

b **Give the meaning of the announcements on the right in French.**

Flight LH …	is on time
Flight from Jamaika	has been cancelled
Flight BA 244	is delayed by 30 min
Flight US 894	has landed
Flight LH 905	now departing

c **Imagine you are a traveller who has just arrived from Lisbon with TAP airlines. Listen to the following announcements and decide whether they are important to you.**

B2 Checking in at the hotel

a **Complete the following dialogue using words from the box below.**

major – breakfast – rent – stay – floor – available – shower – gym – reservation – call

b **Read the dialogue in pairs.**

Chris: Hi, we have a ① for two rooms, one with bath and one with ②.

Receptionist: Welcome to the Royal Excelsior Hotel. May I have your names please?

Chris: This is Valerie Piskei and my name is Chris Fehily.

Receptionist: Let me see. Yes, here we are, Ms Piskei has room 457 and you, sir, have room 476; the two rooms are on the same ③.

Valerie: Could you tell us what time ④ is served?

Receptionist: From 6 a.m. until 10. Room service is also ⑤.

Valerie: Right. I'd like to have a wake-up ⑥ at 7 o'clock tomorrow morning.

Receptionist: I'll note that for room 457. Shall we call you as well, sir?

Chris: No, thank you, I wake up early wherever I am. I'll go straight to the ⑦ to work out before breakfast.

Valerie: We'd like to ⑧ a car for a few days after our business is completed. Do Budget Wheels have a representative here?

Receptionist: We have a car rental desk just across the hall with representatives from all the ⑨ companies. They're open from 9 to 5 every day.

Valerie: OK, thanks.

Receptionist: You're welcome. Enjoy your ⑩ with us!

B3 At the car rental desk

a Study the following table and find out which rate would be favourable for Valerie and Chris. They are going to rent a car over the weekend to do a little sightseeing in and around Philadelphia.

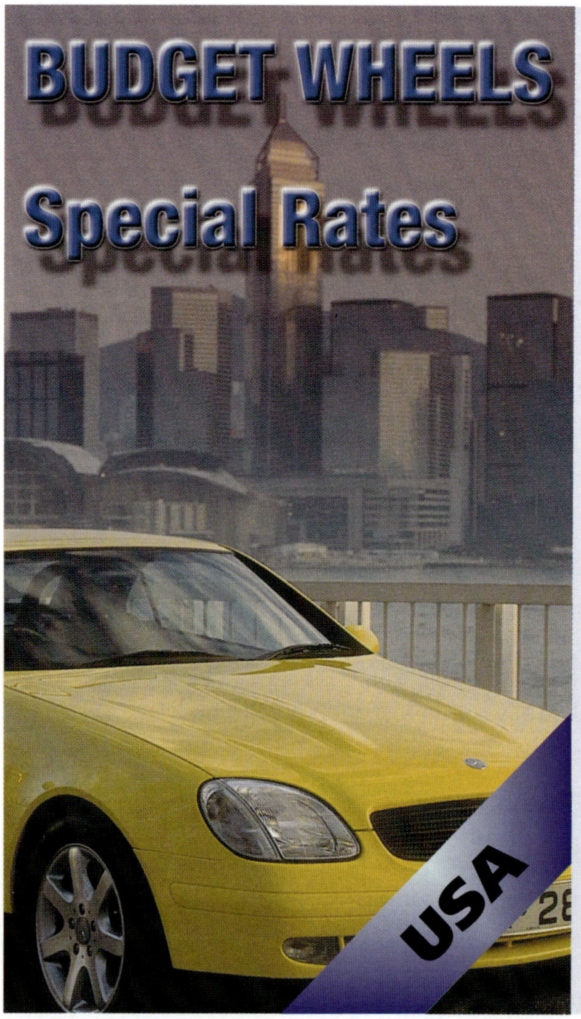

TYPE OF CAR	RATES PER DAY	PICK-UP TIME	RETURN TIME
Compact (A)	$39.75	10 a.m.	no later than 10 a.m. the following day
Mid-size (B)	$58.95 flat rate	any time	any time the following day
Full-size (C)	$65.75	any time	any time

Special Weekend Rates Friday evening 6.00 p.m. until Monday morning 8.00 a.m.

Compact (A)	**$49.75**	**Friday evening, no later than 6 p.m.**	**Monday morning 8 a.m.**
Mid-size	**$65.75**	**see type A**	**see type A**
Full-size	**$69.99**	**see type A**	**see type A**

Note all rates include free mileage!

Type A cars to be returned only to the original car rental office; types B and C may be returned to any branch within Philadelphia and the Phila Intl Airport ($25 extra charge).

Mid-size: holds 2 large and 2 medium-sized people, 4-cylinder engine, automatic transmission, air conditioning.
Full-size: holds 2 large, 2 medium-sized people and 1 small person, 6-cylinder engine, cruise-control, automatic transmission, air conditioning.

b Which car would you choose for 4 adults with baggage? You want to pick up the car at the car rental office in the city center and leave it at the airport. Give reasons for your choice.

c You need a car from 4 p.m. until 6 p.m. the following day. Which one is the best offer? Why?

C Cultural aspects

C1 A walk around the historic centre of Philadelphia

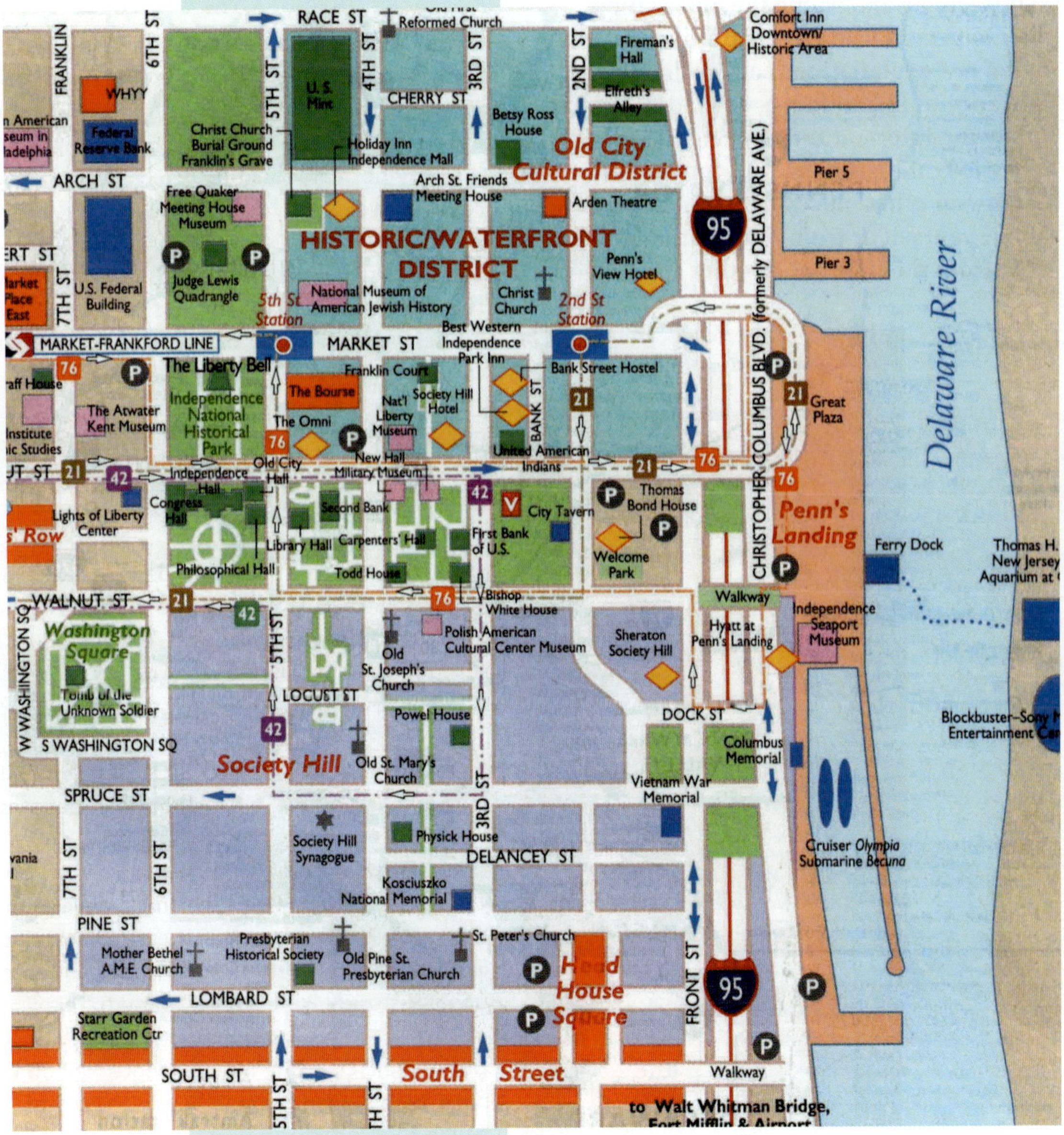

a **Study the map of Philadelphia's historic square mile. Determine where you can find**

- **parking space** Ⓟ
- **museums**
- **hotels** ◆
- **historic buildings**

Example:
There are two parking lots near Penn's Landing.

b Valerie and Chris have an afternoon free, so they want to walk through the historic part of the city. They could either go by car or take the Septa train. **Make suggestions for a sightseeing walk. Name the places they should visit in a convenient order and also add a few explanations concerning the importance of the buildings or places. Use the following information on some of the historic buildings. Feel free, however, to recommend other places that you may know about yourself or from information gathered from the Internet or other sources.**

Independence Hall

Betsy Ross

Philadelphia
– Some historic sights in alphabetical order

Betsy Ross House on Arch Street: Betsy Ross is said to have designed and sewn the first American flag. The museum not only informs the visitor about the history of the flag, it also provides an insight into the way normal working citizens lived during the colonial age.

Carpenters Hall on Chestnut Street: This is where the *First Continental Congress* met in 1774. The house was built by the oldest trade guild in America. Of course old tools are also exhibited.

City Tavern on 2nd Street: An old inn where waiters and waitresses in historic costumes serve meals and drinks. A perfect place to take a break.

Elfreth's Alley (off 2nd Street, between Arch Street and Race Street): If you want to see and feel how it must have been in colonial Philadelphia, walk through this narrow street with its row of old, original houses.

Independence Hall on Chestnut Street: This building saw and heard the most important debates between the young revolutionaries, and it was here that the Declaration of Independence was signed in 1776. In 1787 the Constitutional Convention was held at Independence Hall. The building is located in the small and very peaceful Independence National Historical Park, a lovely place to relax.

Liberty Bell: Opposite Independence Hall on Chestnut Street, a small building houses the most famous bell in the United States. It was cast in London and brought to Philadelphia as early as 1752 to celebrate the 50th anniversary of Pennsylvania's charter of privileges. The crack in the bell that everybody wants to see and even to feel was caused by the first ringing of the bell. It was then recast a second and a third time by two Philadelphians, John Stow and John Pass, whose names are also inscribed on the bell.

C2 Eating out

At the end of their walk through the historic square mile of Philadelphia, Valerie and Chris want to have a nice meal in a typical Philadelphia restaurant.

a Look at the menu and choose your personal meal. Imagine you have about $40 to spend.

b Some menus use different names for the different courses, such as Hors d'oeuvre for First Courses, Entrees (sometimes spelled Entrées) for Main Courses. Also, many menus, especially in Chinese restaurants, subdivide the main courses into

- Beef /Veal
- Pork
- Lamb
- Poultry
- Game
- Fish
- Seafood
- Vegetarian

Find examples for the different subdivisions of the main course in the following menu. Use a dictionary if necessary.

Chef de Cuisine ~ William Legrand

Our Menu

First Course

Tavern Crab Cakes served with herbed remoulade **$8.95**

Giant Cornmeal Fried Oysters pickapeppa remoulade **$6.95**

Smoked Pennsylvania Brook Trout creamed horseradish **$6.95**

Stuffed Shrimp jumbo baked with crabmeat, sherry-cream sauce **$7.95**

Peppered Smoked Salmon served on a potato cake **$8.25**

Soups

West Indies Pepperpot Soup an original colonial soup **$4.25**

Tavern Soup our chef's daily special **$3.95**

Salads

Tavern Country Salad fresh asparagus, cucumber, garden vegetables with fresh herb vinaigrette **$5.50**

Romaine & Roquefort Salad crisp romaine topped with Roquefort cheese, fresh bacon bits & burgundy Dijon mustard vinaigrette **$6.95**

Entrees

Roasted Duckling glazed with pecan honey, served with a peach chutney **$21.95**

Roast Prime Rib of Beef served on an old-fashioned Yorkshire pancake with potato and seasonal vegetable **$23.95**

Medallions of Venison served with a mushroom sauce and vegetable of the day **$23.95**

Chicken Breast Madeira marinated in fresh herbs, sautéed and topped with a madeira-mushroom sauce, colonial wild rice and vegetable of the day **$19.95**

Thai Green Chicken Curry, Sweet Potatoes and Vegetables in a Thai Green curry sauce made with lemongrass, coconut milk and fresh green chillies **$9.95**

Rack of Lamb roasted with Dijon mustard and black walnuts, red roasted potatoes and country squash **$25.95**

Veggie Omelette with mushrooms, avocado, spinach, and cheddar. Served with garden salad. **$12.95**

Tavern Lobster Pie fresh Maine lobster sautéed with shrimp, mushrooms and shallots, baked in a pewter casserole with puff pastry, complemented with seasonal vegetables

Desserts

Baklava Layers of Filo Pastry, Hazelnuts and Pistachios in Spiced Syrup

Creme Caramel Custard Style Vanilla Soya Dessert Topped with Pure Caramel

Beverages

Cold Mineral water $3.95 – Cranberry Juice, Orange Juice $2.95 – Soft drinks $2.95 Iced Tea, Lemonade $1.50 – Draft Beer $4.25

Hot Coffee, Black Tea $1.50 – Herb Tea $1.50 Hot Chocolate $1.95

Wines Domestic: New York State, California, Washington State, glass $4.25 Imported: Italy, France and Spain, glass $4.75

Spirits Whisky: bourbon, rye, scotch $4.95 Gin, Rum $4.95 Cordials $5.25

c **Check the menu on the previous page and find the English terms for the following French words. Use a dictionary if necessary.**

blanc de poulet – épinards – crèpe – crevette – carré d'agneau – champignon – citronelle – noix – homard – caneton – concombre – boissons non alcoolisées – bière pression

d Chris doesn't like spicy food, he hates mushrooms and he is allergic to seafood. **Find a suitable 3-course meal for him.**

e Valerie is a vegetarian; she also doesn't drink alcohol. **Find suitable meals for her.**

f In America (and in other countries as well) you usually do not simply walk into a restaurant and sit down at the next empty table. A waiter or a waitress will help you as shown in the following jumbled sentences.
Put the sentences into the correct order.

1. After giving you some time to choose, he or she will take your orders.
2. When you are comfortably seated, he or she will lay the table for you, if it is not already set up, and bring you the menus.
3. A waiter or waitress will soon come to welcome you and ask how many people are in your party.
4. Waiters and waitresses are generally very friendly and helpful. After all, their service is not included in the meals. You are expected to leave a tip amounting to about 15% of the check on the table, that is, if you are pleased with the service.
5. You wait at the sign or near the entrance.
6. Then he or she will offer you a table and lead you to it.

g **Put the following phrases in the right order.**

What can you recommend?

Is it very spicy/hot?
Is it with meat?

I think I'll have the steak.
I like it rare/ medium/well-done.

Could I have the menu, please?

Could I have some more salt/ pepper/ wine /water …?

Could I have the bill, please?
Can I pay by …card?

D Grammar revision: If-clauses (Conditionals)

See Grammar Overview in the appendix.

a Put in the correct form of the verbs in brackets (*if* clauses).

1. If they (to know) what to do next they would be more relaxed.
2. If they really needed advice they (to call) Richard Shanks right away.
3. Chris said, "I'm sure we won't have any problems with our business partners if we (to be) open about our plans."
4. "That's true," Val said, "but we would have had fewer problems if we (to contact) Richard Shanks earlier."
5. "If we had known about the many ifs we (not/to start) the business over here at all," Chris said.
6. "Well anyway, if this (to turn out) to be a success we will have a real bonus at home."

b Give the German meaning of the following sentences.

1. If we get caught in a traffic jam we'll miss our flight.
2. We wouldn't have to worry now if we had checked out earlier.
3. When would you arrive in Chicago if you took the flight with a stopover in Pittsburgh?
4. What will you buy for your mother if you have time to shop in the duty-free store?
5. Where does your friend usually eat dinner if she has to stay in Chicago?
6. If she likes a restaurant she usually goes back again.

c Translate into English.

1. Si le projet aboutit, nous aurons une raison de célébrer.
2. Nous aurions certainement fini plus tôt si nous avions été mieux préparés.
3. Quel genre de travail aimerais-tu faire si tu avais l'opportunité de travailler en Amérique ?
4. Si je ne suis pas au bureau un vendredi soir, on peut normalement me joindre chez mon amie.
5. Si tu n'arrives vraiment pas à me joindre, envoie-moi un courriel.

Unit 10

Finding a job in Europe

In the course of the past ten years the world economy has changed tremendously. Most large companies are global players now; they produce and sell their goods in many countries all over the world.

The enormous advances in information technologies have enabled national companies to introduce completely new ways of working. Manufacturing often takes place in countries where production is cheap, while the distribution of goods is carried out from the headquarters in the leading industrialised nations. The effects of globalisation are without parallel in human history. Apart from the United States of America, these changes can be felt especially in Europe.

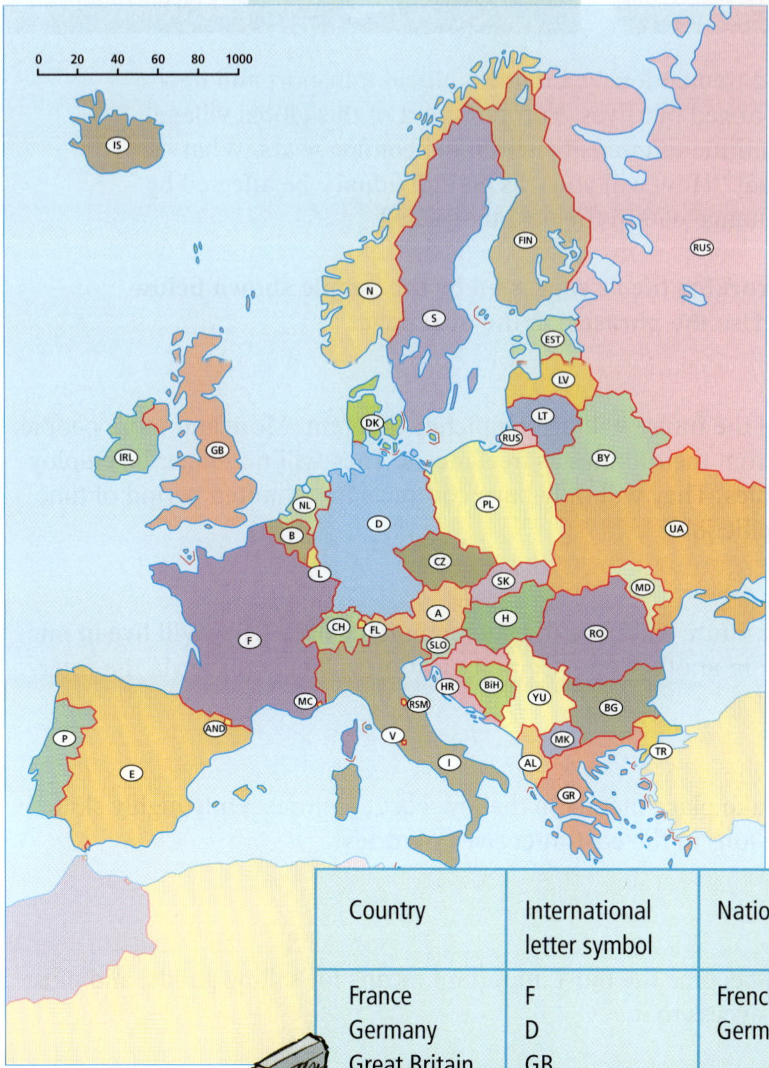

- European countries with the same rights as members of the European Union, according to the European Economic Area Agreement, are Iceland, Liechtenstein and Norway.
  These countries are not members of the European Union. However, the European Economic Area Agreement grants nationals of these countries the same rights to enter, live and work in a country of the European Union.

List the European countries according to the international letter symbols in the map as follows:

Example:

Country	International letter symbol	Nationality	Language(s)
France Germany Great Britain	F D GB	French German …	French German …

A Our workplace is changing

The mobility of people has increased tremendously. Limitless global transport and new methods of communication have changed our lives. Europe is part of this global village. The process of globalisation will continue to intensify within the coming years. What effects will this have on the individual? How will you – as an individual – be affected by this process? How might your life change within the next ten years?

a Discuss the possible effects on working life as suggested by the people shown below and describe your expectations. Use the phrases on the next page.

Björn:
The workplace of the future will be completely different. More and more people will be working from their homes as freelancers. They will no longer be employed by big companies. They will only have contracts for a limited period of time to perform a specific job.

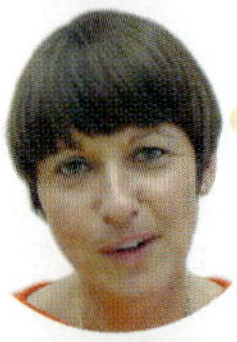

Hélène:
Employees of the future will have to be a lot more flexible. They will live in one country but work in another one. Communication will take place via the Internet.

Marco:
Production will take place mainly in the low-cost countries. Only highly skilled workers will find jobs in the cost-intensive countries.

Nadja:
The Internet will become the most important means for selling goods, and most people will have access to it.

Example:

I agree with Björn as far as the workplace is concerned. But I am not convinced that more and more people will be working as freelancers. I think that the majority will still be employed by big companies.

b Describe how globalisation will affect you. Use the arguments listed below:

- Foreign languages will be more important.
- People's mobility will increase.
- Working with computers will be the norm.
- Teleworking will be widespread.
- Life-long learning will be essential.
- The quality of goods will improve.
- Living and working in foreign countries will be more important.
- Communication will take place via the Internet.

Phrases: Discussions I

From my point of view, ...
As far as I am concerned ...
I am absolutely convinced that ...
We will most certainly ...
It is a fact that ...
I think everybody agrees that ...
I would say ...
Taking these arguments into consideration, ...
To sum up, ...

Example:

From my point of view, foreign languages will be more important in our future economy.

B Working in Europe

B1 Working in the UK

As a reporter for a students' magazine, you have to write an article about working and living in Britain as a European national.

a To obtain background information on the issue, a reporter (Student A) interviews an official of the British Home Office (Student B). Read the dialogue in pairs.

Interviewer: "Do I have the right of residence in the United Kingdom as a national of the European Union?"

Official: "The European Community law grants nationals of the European Community the right to live and work in the United Kingdom."

Interviewer: ”What rights do I have if I want to work in the United Kingdom?”

Official: “You have the right to accept offers of employment or to be self-employed – for example, you can set up your own business. You don’t need a work permit.”

Interviewer: “Do I need to apply for a residence permit or register with the police?”

Official: “As a European national it is not necessary. You are allowed to stay as long as you wish without obtaining a residence permit. There is no need to register with the police.”

Interviewer: “Can I live in the United Kingdom if I am not working?”

Official: “If you have sufficient funds to support yourself throughout your stay without help from the British social security system, there is no problem. You have the same rights as in your home country.”

Interviewer: “Can I work in the United Kingdom if I am studying?”

Official: “No problem. You can work in the United Kingdom during and after your studies.”

Interviewer: “What about my family?”

Official: “The members of your family may join you at any time. During your studies your family members are restricted to your spouse and your dependent children only.”

Interviewer: “Can the members of my family work in the United Kingdom?”

Official: “Certainly. European law applies to them in the same way as to you. They do not need work permits.”

Interviewer: “Do the same laws apply to all the other countries of the European Union as well?”

Official: “As this is common European law, you can live and work in all the countries of the European Union.”

Interviewer: “Thank you for your help. It is good to know that I have the same rights in the United Kingdom as in my home country.”

Official: “Well, that is part of our common European idea.”

b Write a short article for a magazine which contains the most important information about working and staying in the United Kingdom as a national of the European Union. Use the information from the interview and start like this:

”Working and living in Britain – no longer a problem”

The “United States of Europe” is no longer a distant dream, as Europe is growing together. European Community law guarantees the same rights to all nationals of the European Union. The right to work and live in Britain has become much easier in recent years.

Go on.

c Apart from nationals of the European Union, similar rules apply to some other groups of people who wish to work and live in Britain. **Find out which people do not need work permits and have similar conditions. The necessary information is available on the homepage of the Department for Employment and Education (DfEE).**

d Why do these people have the same rights as nationals of the European Union?

B2 Telework – Facts and figures

Telework has started to boom in Europe

The Lisbon Summit in March 2000 marked the beginning of a new phase of European co-operation in the shift to a digital, knowledge-based economy. The European governments agreed to turn Europe into "the most competitive and dynamic knowledge-based economy, capable of sustained economic growth with more and better jobs and greater social cohesion." It was agreed to develop telework continually as Europe moves further to a knowledge-based economy on the way to a global information society.
While in 1994 just a few million teleworkers were identified in Europe, the figure has risen to around 10 million in 2000. The range of telework penetration across European countries is enormous, Spain finding itself at the tail end, with 2.8% of its employees doing telework and Finland at the top with almost 17% teleworkers.

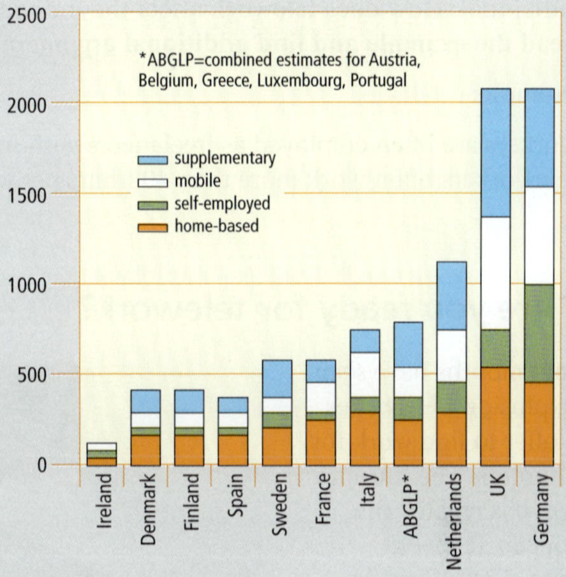

Telework in Europe 1999
Number of teleworkers by country, 1999

a Answer the following questions on the text and the diagram above.

1. What was the main result of the Lisbon Summit with respect to telework?
2. What results do the European leaders expect by promoting telework?
3. Which three countries have the highest percentage of telework in Europe?
4. Which three countries are at the bottom of this list?
5. Which country has the greatest number of home-based teleworkers?
6. Which country has the greatest number of self-employed teleworkers?

b Work in pairs. List the advantages of telework for companies and individuals.

Example:

Advantages for businesses	Advantages for the individual
• teleworker works in his/her home office: fewer offices are needed • higher productivity • improved and much more open communication • decrease in hierarchy • higher motivation of staff • reduced costs • …	• time management: the teleworker can decide when and how long he/she wants to work • new jobs offered on the market • increased responsibility • more freedom • … • … • …

c **Prepare a presentation on the advantages of telework. Work in groups of 4 students. Before starting, collect the arguments of all the students at the board or with the help of a flip chart (if possible, use a computer for your presentation).**

d **Some groups, e.g. the trade unions, fear the effects telework may have on the individual. How does telework affect the individual in a negative way? Read the example and find additional arguments.**

Example:

Workers are often employed as freelancers without any social security.
Teleworkers often work more than 40 hours per week.
...

B3 Are you ready for telework?

Recent months have seen an explosion of websites that offer to find work for job-hunters. The European Union has created the European Telework Online (ETO) Website which will help you to find information about telework.

a **Search the Internet for information about:**

- **Telework in general**
- **Current numbers of teleworkers in Europe**
- **Online services for finding work**

b **Answer the following question after you have searched the Internet:**

What is the general aim of the European Telework Online service?

c **Describe one job offer in the field of teleworking in detail.**

d **Find the ETO's definition of telework and summarise it in your own words.**

e **Are you ready to become a teleworker? Describe a job in teleworking that would interest you.**

C Jobs, job application and interview

C1 Job application

TELEWORK:
MULTIMEDIA DESIGNER

Do you know what advertising on the web is all about?
You will be responsible for the development of the design and its successful implementation using graphic software and programming.
You can work from anywhere, and define your own hours.
Essential Requirements: Knowledge of current media technologies and experience with their application (animation skills and knowledge of video and sound editing a plus). You must be able to communicate ideas graphically and to handle projects individually.
You are used to meeting deadlines.
Salary: Negotiable

Letter and CV to:
Margret Tuson
MultiWeb Ltd
103 Sinclair Road
London W 14
phone: 020 7324 9984
fax: 020 7324 9999
email: m.tuson@multiweb.com

Technical Project Manager – Europe

We are looking for someone with the ability to manage projects from conception to completion. You will be planning, scheduling and overseeing the implementation of different technical projects in Europe. Good knowledge of computer hardware and networks is essential, as you will be working closely with our software development teams.

Fluent English and a good knowledge of French (and possibly Spanish) are essential.
Salary is negotiable. Fringe benefits + company car.

Contact:

Michael Farley

m.farley@eurojob.de
EUROJOB International
Phone: +49-911-54053-121
FAX: +49-911-541-7759

MANUFACTURER OF MACHINE TOOLS AND POWER TOOLS IN SWITZERLAND REQUIRES

2 Electronics / Mechatronics Assistants

with practical thinking and hands-on approach.

You will work in an international team with mechanical and electrical/electronic design engineers.
Qualifications: 1-2 years experience in Electronics/Mechatronics (or equivalent)
Basic CAD skills

Languages required: Fluent French, good English and German

Applications to:
Swisstools
rue du Lac 12-18
1077 Pully
Switzerland

BERLIN-BASED COMPANY EMPLOYS

3 QUALITY ENGINEERS IN NEW BRANCH IN BIRMINGHAM

Salary Neg. + Benefits package

Experienced Quality Engineers, preferably from an electronics or automotive background.
Must be familiar with Quality systems ISO 9000 and 14001.
Languages: Fluent English and French

Apply to:
Cartronics Ltd.
134 Alum Rock Road
Wood End, Birmingham
West Midlands. 0121 312 2616

a Read the advertisements on page 153 and above and find the English translations for the following French words and expressions.

1. responsable de
2. logiciels
3. exigeances
4. actuel
5. traiter
6. dates limites / délais
7. achèvement
8. contrôler / superviser
9. avantages en nature
10. de préférence

b Which advertisement appeals to you most? Give reasons.

c Use the following information to write an application letter. Think about capitalisation, punctuation, paragraphs and layout.

dear mr white I would like to apply for the job of technical consultant advertised in *Science and Technology* since completing my education I have been working for Boltex S.A. a car component supplier where I am technical assistant to the sales director my duties include quality control and budget planning I feel that my technical knowledge and in particular my language abilities in english and spanish are ideally suited to a position in an internationally operating company thank you for considering my application if you need any further information please write or call me yours sincerely carsten berger

d Study the advertisements again and write an application for one of the jobs. Use fictitious data and follow the sample application you have written above.

e Read the information in the INFO-BOX on the next page and write a similar CV using your own or fictitious data.

INFO-BOX: Curriculum vitae (CV) or résumé (in the USA)

If you want to apply for a job you have to write a CV which contains your personal history in short. A CV should include the following information:

Example:

Name	Janet Rush
Address	11 Collville Gardens, London W 12, GB
Place of birth	Brighton
Nationality	British
Age	25
Education	2000 - 2001 University of Reading, Post Graduate Studies
	1997 - 2000 University of Reading, B.A. in Electronics
	1990 - 1997 Brighton Comprehensive School
Languages	English and French
Previous experience	Dumarchand S.A., Créteil; 6 months as trainee in R & D.
Interests	Travel, CAD, art, skiing

Tips on writing a CV

When writing a CV always use a word processor. For a good presentation it is necessary to think of the following rules:

- When you apply in GB the CV should not be longer than 1-2 pages.
- Use good-quality paper.
- Point out your own strengths and achievements.
- Include a photo (photo optional in GB).
- Include written references (not required in GB).
- Never send a CV without a covering letter which includes your address and the date.
- Start with your most recent job when describing your education and previous experience.
- Check your spelling and ask a friend to double-check it.
- Don't include your salary, unless requested.
- Don't ask the company to send back your CV – most companies keep it 'on file'.
- Keep a copy of your own CV and look at it again before your interview.

Comparision of French school types and their British equivalents

Before applying for a job in a foreign country, it is important to find out about the differences in the way applications are written. Furthermore, it is often difficult or even impossible to find the equivalent of your qualifications in other countries. If there is no good translation or equivalent you can always describe how long you studied and what subjects you took.

The following table gives some French educational terms and shows how they can be explained in English:

French system	English system
école maternelle/primaire	primary school (elementary school in USA)
collège	secondary school (junior high school in USA)
lycée	secondary school (senior high school in USA)
lycée d'enseignement professionnel	technical school
institut de technologie	technical college
formation en alternance	sandwich course

C2 Interviews

a Work in groups. Prepare your own answers to the following "tough" questions for a job of your choice.

"Tough" questions:	Tips for your answers:
1. Tell me about yourself.	Keep answers short; use CV summary.
2. What do you know about our company?	Find out about company, including products or services, size, number of employees, company history and philosophy.
3. Why do you want to work for us?	Focus on needs of company; you have knowledge and experience to make a contribution to meet company needs.
4. Why do you want to leave your present job?	New challenge, no career prospects, location, security.
5. What can you do for us that someone else can't?	Qualifications, specific experience in similar position, language skills.
6. Do you prefer working with others or alone?	Whatever is suitable for the job.
7. What are you good at doing?	Focus on skills that are required for the job.
8. What do you dislike doing?	Name only one minor thing.
9. What about the job offered do you find the most / least attractive?	List attractive factors, but only one minor unattractive factor.
10. What do you hope to be doing in 3 (5) years?	Say that you are confident about meeting the expectations of the employer and that you feel you can take on more responsibility in the near future.

**b Read the following list of tips for interviews.
Three of the tips are false and inappropriate. Find them.**

1. Arrive on time or a few minutes early. Arriving late for a job interview creates a very poor impression.
2. Smile and shake hands firmly.
3. Listen carefully and speak clearly.
4. Don't wait until you are offered a chair. Be sure of yourself and make yourself comfortable.
5. Sit upright and look alert and interested at all times.
6. Focus on your achievements and successes in your present or in former positions.
7. Don't smoke, even if the interviewer does and offers you a cigarette.
Do not chew gum or eat candy.
8. Answer questions with "yes" or "no" whenever possible.
9. Answer questions truthfully.
10. Never make critical or derogatory remarks about your present or former employers.
11. Don't inquire about salary, vacations, etc. unless you are sure the employer is interested in hiring you.
12. Never ask questions about the history of the company.

**c Work in groups of three and role play an interview. Student A is the candidate.
Students B and C are the interviewers. Take it in turns to play the candidate.**

Interviewer

What made you apply for … ?
Could you tell me about … ?
What do you know about … ?
I'd also like to know … ?
What are your qualifications …?
What are your strengths / weaknesses?
Have you ever done / worked / used … ?
Are you familiar with …?
Do you have any questions you would like to ask?

Candidate

I saw your advert in …
I am interested in …
I think that …
I studied ... in ...
I worked for … in the … department.
I worked as a …
My main tasks were …
I was responsible for …

C3 The Green Card – a controversial issue

In many industrialised countries there is a lack of specialists, especially in the new IT branches. Offering 'Green Cards', i.e. work permits to foreign specialists, is not new. This principle has been applied in the United States of America for a long time. Some European countries have introduced similar schemes to recruit foreign specialists. This issue has become the subject of controversial discussions.

- **Role play: Panel discussion**
  Work in groups of six and follow these steps:
  - Rearrange your classroom for the panel discussion.
  - The panel discussion begins with an opening statement from each panel member (time: 2 minutes). To prepare this panel statement, each panel member is supported by at least two more students. Group work for preparation: 10 minutes.
  - The chairman opens the debate by introducing the topic and the participants to the audience. He decides on the order of the speakers during the opening statement.
  - During these statements the speakers may not be interrupted; each panel member should take notes during the other speakers' arguments (necessary for the debate afterwards).
  - After all statements have been made, a debate is started; the chairman is responsible for the discussion and the order of speakers.
  - The panel discussion should last about 10 to 15 minutes. The chairman closes the debate.
  - The remaining class members, i.e. the audience, should take notes in order to write a report at the end of the discussion.
  - If you have the facilities at college, a video of the panel discussion can be made to discuss the TV show in class.
  - Each group member must take on the role of one of the following characters. Use the phrases given to start the discussion.

Jennifer Bates (an industrialist who wishes to employ foreign specialists)

Due to a shortfall in specialists in the domestic labour market ... we have been forced to ....
There are few alternatives other than ...
Well-qualified people are hard to find ...
The Green Card is an optimal solution for us because ...

Bill Walker (an educational specialist who wishes to improve training at colleges and therefore rejects the idea of the Green Card)

The introduction of the Green Card proves that education ...
As a short-term solution the Green Card ...
However, in the long term ...
We should be investing in our own training colleges, not ...
Education and training have been neglected by successive governments ...

Susan Morris (an unemployed computer specialist who certainly does not like the idea)

I am totally against the idea ...
The government should ...
The level of unemployment is high enough ...
There is enough competition in the labour market without ...

Gurjit Singh Bedi (an Indian IT specialist who would like to apply for a Green Card)

The Green Card is a great opportunity ...
Indian IT specialists are highly skilled ...
It is a practical solution ...

Bruce West (a politician who was responsible for the introduction of the Green Card)

Let me just make 3 points. Firstly, the development of IT professions ...
Secondly, ...
As far as the government is concerned ...
We need practical solutions fast, and ...

Harriet Smith (in charge of the panel discussion as chairwoman)

Right. Let's start our discussion.
We are here to talk about ...
Mr/Ms ... could you comment on that?
One at a time, please!
Would you like to add something to this point, Mr/Ms ...?
I think it is agreed that ...
Now let's move on to (the problem of ...)
To summarise the discussion, it can be said that ...

Phrases: Discussions II

In my experience, ...
In my opinion, ...
I see your point, but ...
Exactly!
I don't quite agree here. As I see it, ...
Sorry, could I just finish?
If I could just interrupt here, ...
I think you'll find that ...
I'm afraid I can't agree ...
Looking at the statistics ...
This proves beyond doubt that ...

D Grammar revision: Future tense

See Grammar Overview in the appendix.

a Choose the correct tense from the verbs in brackets.

1. (I go / I'm going) to Brussels next week.
2. Our train (is leaving / leaves) at 8 o'clock tomorrow morning.
3. How long (will you stay / will you be staying) in Philadelphia?
4. Our office (is going to / opens) at 9 a.m.
5. Wait a minute! I ('m going to / 'll help) you.
6. I suppose Jane's application (will be / is going to be) accepted.

b Complete the sentences. Use the most suitable type of future tense.

London Telework has invited all their new employees to dinner tonight.

"I can assure you that I (1. to do) my best to make a good impression.
I (2. to wear) my best suit and I think I (3. to have) a haircut, too, before I go.
My train (4. to leave) at 5.30 p.m. We (5. to meet) at the Old Ascot at 7.30 p.m. Dinner (6. to start) at 8.30. While you (7. to watch) the news tonight I (8. to enjoy) excellent food and wine. I suppose dinner (9. to be over) by about 10 p.m. As the last train (10. to depart from) Paddington Station at 10.30 p.m., I expect that I (11. to be) home by 11.30 p.m. So don't wait for me; it (12. to be) a late night. Oh, it's only 3 o'clock in the afternoon and it's getting dark outside. It (13. to rain). I (14. to need) an umbrella. Can you help me find one?"

c Translate into English. Use suitable forms of future tense.

1. S'il est promu, il aura sans doute un plus grand bureau.
2. Je vais postuler pour un poste à l'étranger dès que possible.
3. Nos clients japonais vont visiter l'usine demain.
4. Je suis sûr que tu prendras la bonne décision.
5. Attends ! Je vais t'aider !
6. Je pense que la demande en spécialistes informatiques va augmenter.
7. Mon avion décolle à 16h30.
8. Vas-tu postuler ?
9. Nous allons remplacer nos vieilles machines. La décision a été prise à la réunion d'hier.
10. Demain, à peu près à la même heure, je serai dans l'avion pour Londres.

Reading Files

File 1 Civil engineering

Civil engineering can be defined as the planning, design, construction and maintenance of structures, which may include the altering of the landscape and manipulation of nature in order to meet human requirements. There are many subdivisions of civil engineering, some of the more common ones being transportation, for example the creation of roads and railways; hydraulics, including river control and the provision of sanitation; and structures, which is concerned with buildings, bridges and tunnels.

Design and function

In each area of civil engineering the relationship between design and function always has to be taken into account. The design and form of a structure will depend on the intended use, as each different structure will be subject to different loads, for example the weight of books in a library, or the difference in the wind loads imposed on low and high-rise buildings. In order to design a structure which fulfils its function optimally, a civil engineer will have to consider all these aspects. Buildings can be used for residential, storage or business administration purposes, so the civil engineer will produce plans for houses or apartment blocks, warehouses or office buildings, all of which have very different design demands.

Residential buildings are designed with separate rooms for different purposes, such as the kitchen for cooking, the bathroom for washing etc. The external walls and the roof form the outer frame, with the internal space being divided up by walls and intermediate floors. Internal walls which carry some of the structural weight of the building are called load-bearing walls. Non-load-bearing walls are called partition walls. Easy and safe access from one part of the building to another should be included in design plans, as should other important factors such as the provision of light, ventilation and warmth. As a result, further building services are required to provide lighting, heating, connection to a water supply and drainage.

Storage buildings, such as warehouses, require as much internal space as possible and consequently large open spaces with high ceilings are usually preferred. An important design consideration is access for large vehicles to deliver and collect goods, and enough free space for fork-lift trucks to move and stack goods inside. In addition special provisions are needed to reduce fire risk.

Yet another set of considerations has to be taken into account when planning office structure. Allowances need to be made for a large number of people working in a single office, so there should be plenty of daylight as well as good ventilation and heating systems, as modern office buildings are frequently air-conditioned. The internal design must enable movement from one part of the building to another, so corridors, hallways, lifts, and staircases are obligatory, as are emergency escape routes in case of fire.

Structural forms

The walls have a key function in conventional building construction, because they form the enclosure and transfer all the loads to the ground, resulting in a solid structure. Thick masonry walls have been used for centuries, and more recently mass concrete and brick have been used to form the structure. Nowadays structural steel engineering and reinforced concrete structures also play an enormous role, especially in industrial architecture. Floors must also bear loads and be constructed in such a way as to ensure an even load distribution. For this reason floor spans must not be too great, as this could weaken the load-bearing capacity.

Building elements and materials

What materials are used in structural engineering?
This depends on the building elements. Concrete, steel, bricks and blocks are used for foundations; bricks, blocks and timber for walls. Floors are built with timber, concrete, or sheet materials, and roofs are made out of timber, steel, concrete, clay products, metals and even plastics. Frames are often made out of concrete and steel due to their load-bearing function.

The loads, the forces acting on the building structure, are a combination of dead, imposed, wind and live loads, and must be transmitted safely to the ground. Dead loads result from permanent constructions such as the weight of walls, partitions, floors, roofs and finishes.

Imposed loads depend on the weight of movable partitions and on how these are distributed or concentrated, for example store goods or snow loads on the roof. Wind loads are affected by wind pressure or suction; live loads are produced by the changing size or position of the load, for example traffic moving over a bridge. Full design and structural calculations have to be made if loads exceed certain specified values, such as stability requirements for small buildings of traditional construction.

Foundations

Foundations spread loads over the ground in such a way as to ensure that the building load is supported and the structure itself is stable. The type of foundation and its required strength depend on the weight of the building, its structural form and the type of soil.

The most common type of foundation is the strip foundation, which is normally constructed for load-bearing walls. A trench is excavated to a suitable depth and a concrete base is compacted into the bottom and then built up to ground level with either brick courses, concrete fillings or concrete blocks. Although concrete has a good compressive strength, it is weak in tension. Consequently, steel reinforcing bars are often needed. To protect the steel from corrosion, it is essential to cover it sufficiently in concrete.

Walls

The walls should be set centrally on the foundations, as they fulfil many different functions. They have to carry and transmit loads, to provide security (for example by restricting the spread of fire), to resist the wind and weather, to restrict heat loss – an important environmental aspect – to divide space, to provide acoustic insulation and visual privacy, to prevent damp penetration, etc. The materials used for wall constructions are timber, bricks, blocks and concrete with or without steel reinforcement. The stability of a wall is paramount and depends only partly on the strength of the materials used. Walls are usually thin in relation to their overall size, and this presents a problem for structural stability; i.e. they have to be prevented from buckling, toppling or being pushed sideways. The strength of the materials must be sufficient to carry the imposed loads with a sufficiently large factor of safety. Walls of bricks or blocks are bonded for strength to ensure that vertical joints do not come directly one above the other.

Other elements which increase the stability of a wall are built-in piers and buttresses. A pier is a column of masonry, thicker than the wall itself, which can also be used to support a concentrated load such as a beam. A buttress is built at right angles to the line of the wall; it may be a pier or another wall, but it is often a chimney stack. The ends of the walls should be securely tied into a buttress, while long walls should have an intermediate support.

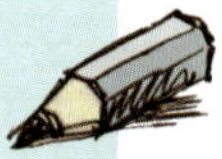

Floors

The construction of floors and roofs provides lateral support for the walls. Floors are required to carry loads including self-weight and to provide fire resistance and sound insulation. The loading on a floor results in a bending moment which is greater for larger spans. The greater the floor span, the deeper the joists must be in order to prevent deflection. Fir, larch, pine, spruce and imported redwood or whitewood are frequently used for flooring, as is concrete. Different shuttering methods and techniques play an important role in the construction of suspended concrete floors unless pre-cast reinforced concrete or pre-stressed concrete are used. The latter provides greater strength and can be used for longer spans.

Roofs

The traditional roof construction is the pitched roof, as illustrated in the picture on the right.

The basic shape of a pitched roof is a triangle. Within this triangle are smaller triangles formed by the hangers. The triangle shape is very stable because it cannot be distorted without squashing or stretching the lengths of the sides.

a Match the building elements with the materials to be used.
Note: Some of the materials can be used for several building elements.

building elements **materials**

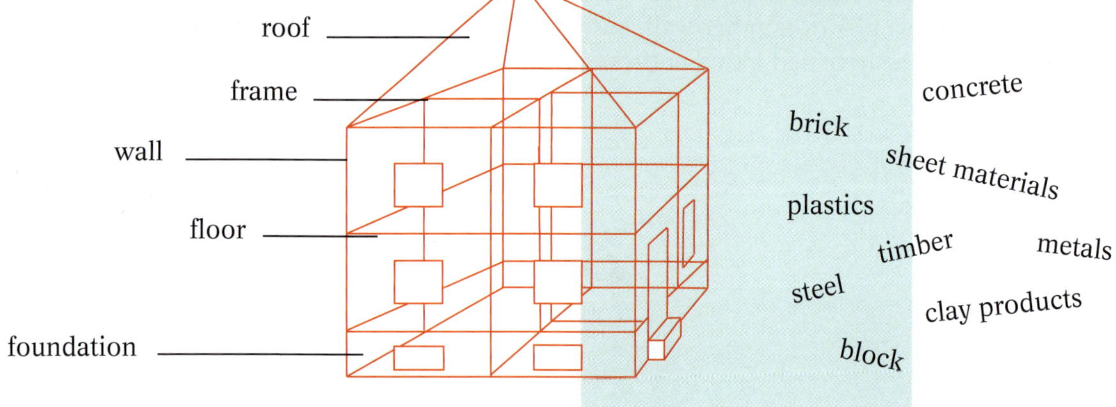

b Read the text on civil engineering again and complete the following sentences using words from the box below.

1. Internal walls which carry no structural weight are called ? .
2. Water supply and ? are two examples of building services.
3. Over the centuries thick ? has been used for solid building.
4. Today, in industrial architecture ? play an enormous role.
5. The forces acting on the ? are called loads.
6. Traffic moving over a bridge produces ? by changing the size or position of the load .
7. The most common type of foundation is the ? which is normally used for ? .
8. Work on a foundation starts with excavating a ? .
9. As ? is weak in tension ? are often needed.
10. The stability of a wall is increased by built-in ? and ? .
11. They are often used to support a concentrated load from a ? , for example.
12. The loading on a floor depends on the ? .
13. Long spans require great strength, which is provided by ? .
14. The traditional roof construction is the ? .

spans – masonry – steel reinforcing bars – load-bearing walls – trench – piers and buttresses – pre-stressed concrete – reinforced concrete structures – partition walls – live loads – beam – shuttering methods – strip foundation – drainage – building structure – span – pitched roof – concrete

File 2 Woodworking

A Professions

There are three main professions that involve working with wood: those of the **carpenter**, the **joiner**, and the **glazier**.

- **Match the correct professions (carpenter; joiner; glazier) with the numbers in the text below.**

The ① works on the structural parts of a house, such as the framework of the outer walls, the heavy beams for roofs and floors, and chipboards. He / She usually works outside. The ②, however, works inside. The wood laths he / she uses are not as rough as the beams, and they are smaller. He / She produces window frames, doors, and furniture. The ③ fits glass panes, for example in window or door frames.

B Materials

What just looks like wood for the layman is an assortment of materials with very different properties and uses for the specialist. Customers prefer their furniture or their wooden ceiling made from real wood. It looks good, smells good and is healthier than many artificial substitutes.

a Match four of the pictures below with the following types of wood: pine – walnut – cherry – maple

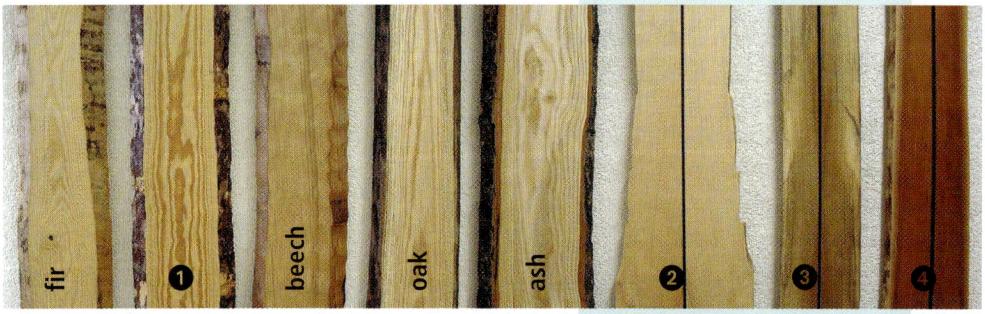

In most cases, however, solid wood is too expensive. So joiners use veneer instead, very thinly cut sheets of real wood glued onto laminated fibre sheets.

b Find out about the production of Oriented Strand Boards (OSBs), chipboards and plywood. Use the Internet or any other source.

C Tools

As in other trades, craftsmen who work with wood have their own special tools. Hand tools such as a hammer, a chisel, a back saw or a Japanese saw, a planer or clamps are still used, but of course joiners and carpenters also use power tools wherever possible.

a Look at the following pictures and give the correct names.

planer – chisel – clamp – keyhole saw – work bench – mallet – brace – file

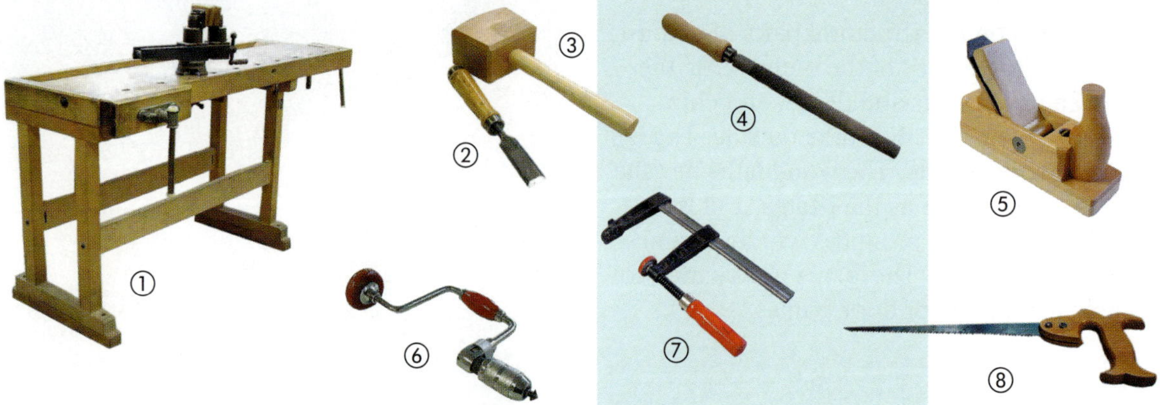

b What can you do with these tools? Complete the text using the names of the tools given on page 166.

With ① you can cut wood to the correct length. With a ② you can take away splinters and edges from a piece of wood, thereby making the surface smooth and even. With a ③ you take away large chips of wood. ④ are needed if you want to hold two pieces of wood firmly together, for example to glue them.

The following pictures show two power tools widely used by joiners:
a jigsaw and an eccentric grinder.

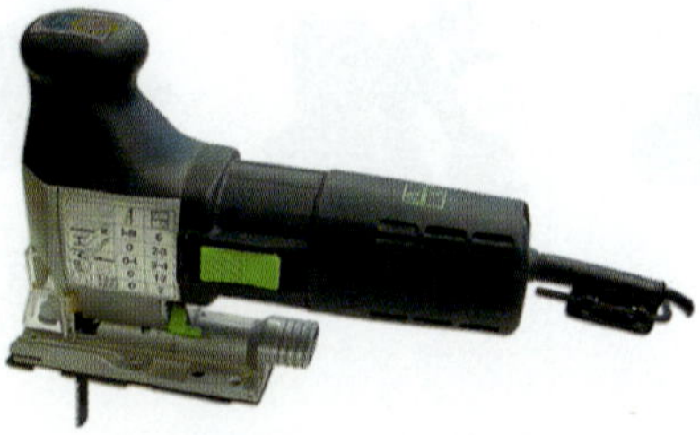

jigsaw

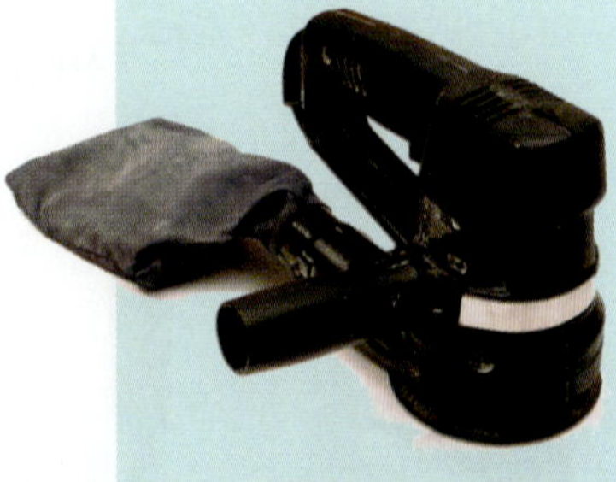

eccentric grinder

c Describe when and where these tools are used. Use a dictionary if necessary.

A ... is used for cutting / grinding / smoothing / sanding / polishing ...

D Work processes: Joining wood

a Complete the following text using words from the box below.

tackers – nails – shrinks – loose – sight – lath – bonds

Nails

A joiner is someone who joins pieces of wood together. There are various ways to do this. One way is to use ① and hammer them through the two pieces of wood, two wooden beams, for example, or to fasten a ② on a beam. Hammering is very fast, but modern machines like strong ③ can also do the work for you. Whereas hammering makes very strong ④ of two pieces of wood for the carpenter, it is not always the best method for the joiner. First of all, nails in a piece of furniture are not a very attractive ⑤, and secondly, nails are not very durable in furniture. Wood usually ⑥ when it dries and thin nails, in particular, can become ⑦.

Screws

A better way of joining wood is with screws.
Wood screws have different threads than machine screws.

b Match the different types of wood screws and their parts as shown below to the French and English terms.

thread
crosshead
cruciforme
tête
star drive
tête ronde
raised head
filetage
tête à six pans creux
shank
head
tige

①

②

③

Glue

Another way of joining wood is by gluing.
Gluing means that you put special glue – usually a milky-white liquid - between two pieces of wood and then clamp them together with screw clamps for a certain period of time to get a very solid and durable bond between the two pieces. But you first have to make sure that the surfaces of the parts that are to be glued together are absolutely flat, smooth and clean, which means free of dust or grease.

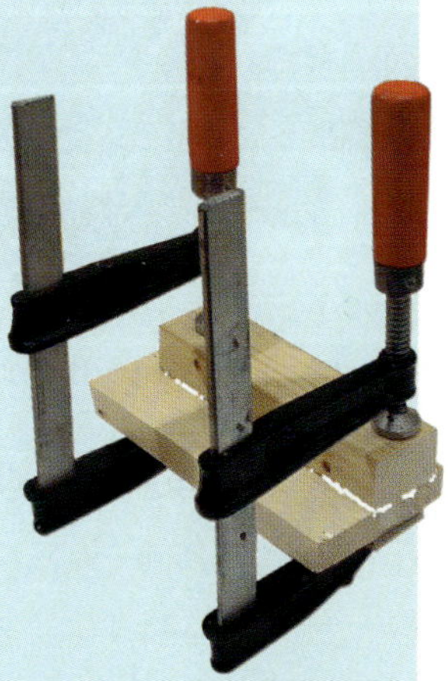

c Determine what different kinds of glues there are in your workshop and describe how they have to be used.

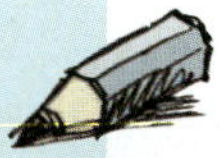

File 3 Laser technology

One of the new technologies that has an amazing range of products and applications is the fascinating laser technology. Lasers can be found in everything from CD players and laser printers to dental drills, high-speed metal-cutting machines and measuring systems. The word “laser” is actually an acronym for “light amplification by stimulated emission of radiation”.
But what is a laser and what makes a laser beam different from a flashlight beam?

How lasers work

A laser is a device which controls the way energized atoms release photons, that is particles of light. Although there are many types of lasers, they all work due to the way light interacts with electrons. The electrons exist at specific energy levels which are characteristic of that particular atom or molecule. The energy levels can be imagined as rings or orbits around a nucleus. The electrons in the outer rings are at higher energy levels than those in the inner rings. If energy such as a flash of light is injected, the electrons can be bumped up to higher energy loads. When an electron drops from an outer to an inner level, "excess" energy is produced and given off as light. The wavelength or colour of this emitted light is accurately related to the released amount of energy. The characteristic laser colour is red, a fact which is connected with an aluminium oxide crystal called "ruby". In this crystal, some of the aluminium atoms have been replaced with chromium atoms, which absorb green and blue light and emit or reflect only red light. So chromium gives ruby its typical red colour and provides the crystal's lasing behaviour.

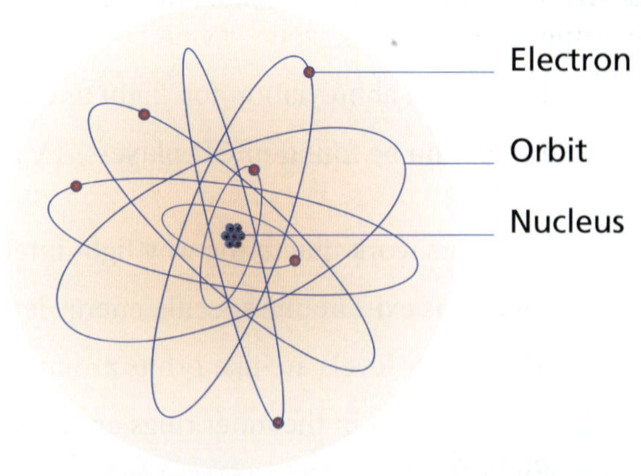

Mirrors at each end of the lasing medium reflect some this ruby light back and forth inside the ruby crystal until the light pulse builds up to high power and drains the energy stored in the crystal in the form of a very intense laser flash which lasts only about 300 millionths of a second.

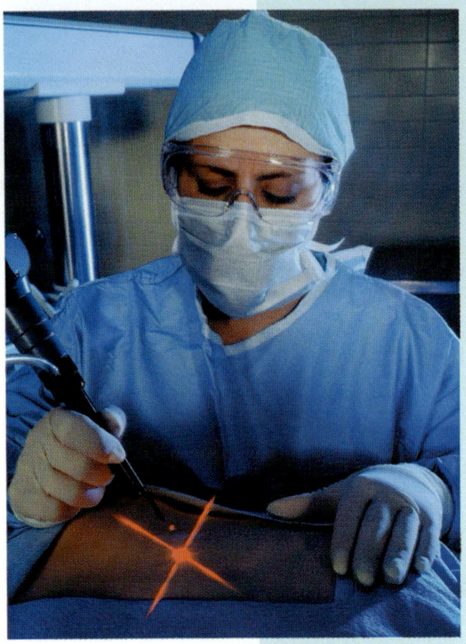

Modern lasers can produce pulses with the power of some billion watts.

Another characteristic of laser light is its coherence, which means that the light can be focused with great precision. There are lasers, for example diode lasers, which are rather weak and are used in pocket lasers to point at things. Or there are lasers used in medicine, for example in plastic surgery or eye surgery.

The CO_2 laser which is used in industry, however, is very powerful and can cut through steel. But this laser is also very dangerous because it emits laser light in the infrared and microwave regions of the spectrum. And because infrared radiation is heat, this laser melts through whatever it is focused upon.

a Are the following statements about the text true or false? Correct the statements if necessary.

1. Laser is an abbreviation for "light and safe electronic radiation".
2. Lasers can be found in CD players, TV sets, power drills, jigsaws and measuring systems.
3. All lasers work due to the way light interacts with electrons.
4. Electrons exist at one specific energy level.
5. The energy levels are like orbits around a nucleus.
6. The electrons in the inner rings are at lower energy levels than those in the outer rings.
7. If energy, such as a flash of light, is injected, the electrons can be bumped up to higher energy loads.
8. The characteristic laser colour is green.
9. Some of the ruby light is reflected back and forth inside the ruby crystal by mirrors at each end of the lasing medium.
10. Modern lasers can produce pulses with the power of some million watts.

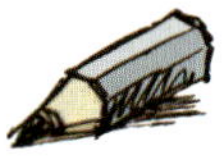

11. Laser light can be focused with great precision.
12. There are lasers, such as the CO_2 laser, which are rather weak and are used in medicine, for example in skin resurfacing or eye surgery.

b Use any sources available and find more applications of laser technology. Then list them under categories according to the different fields of application, e.g. surgery.

File 4 Logistics and transport technology

A Logistics

The term logistics and the idea behind it came originally from the military. Logistics means all the complex operations necessary to arrange for the transport of goods (or people) from point A to point B in a given time, at a price that is acceptable for both the customer and the service supplier.

The American-based *Council of Logistics Management*, a non-profit organization for the logistics business, gives the following definition: "Logistics is that part of the supply-chain process that plans, implements, and controls the efficient, effective flow and storage of goods, services, and related information from the point of origin to the point of consumption in order to meet customers' requirements."

While the concept of logistics is so broad as to include the planning of urban and suburban public transportation systems and the construction of motorways to connect all parts of a country, its principal field is the transportation of goods from one point to another. Some companies have their own logistics department, some hire large national or international specialist companies to transport their cars, machines, machine parts or other products from the place of production to where their customers want them.

Logistics is a prime example of globalisation. In order to transport goods around the globe, communication between the different national and international companies involved must be flawless; freight forwarders, public and private rail and shipping lines, and air carriers must work in perfectly coordinated teams, not to mention the banks and insurance companies that are needed to finance the transportation and to insure the goods against theft, damage, or loss. To get an idea of how well-coordinated such teams must be, think of the daily output of 800–1000 units in an automobile factory, making several thousand every week that have to be delivered to points inland, across borders, and overseas.

a Translate the definition given by the Council of Logistics Management.

b Find out about logistics enterprises (national or international) and report to your class.

c Find synonyms in the text above for the following words or phrases:

1. organise
2. productive without waste
3. perfect
4. the act of stealing

d Find the corresponding noun, verb or adjective in the text for the following words:

1. transportation
2. originate
3. consume
4. coordination
5. financial
6. insurance

B Modes of transport

The simplest logistics task is to load goods on a lorry or truck and transport them to their destination, where they are unloaded again. However, as soon as you have to transport your goods over a greater distance – overseas, for example – the matter becomes more complex: Multiple means of transport are involved. You have trucks to take the goods to a rail freight station; if they are bigger loads you need semitrailers. From these vehicles they are loaded on trains, transported to harbours and onto ships, or taken to airports, where they disappear in the bellies of huge cargo planes. After the cargo has been cleared through customs in the country of their destination, they will be loaded onto another train or truck for the final leg of their journey.
In order to save time, goods are transported in special containers. Their sizes conform to the standards of the International Standardization Organization in Geneva (ISO): therefore, a container loaded in the yard of a French company fits on every container truck, on every train wagon, on cargo planes, and on ships all over the world. Loading and reloading the container is less time-consuming than loading and reloading the goods themselves.
Another method is the roll-on-roll-off system. This means that the loads stay on the trucks and the trucks themselves are loaded onto trains or ferries. The costs of the transportation of the loaded vehicles may be higher, but time and labour are reduced since the containers do not have to be unloaded and then reloaded on arrival.

a Work in groups. Choose different countries and find out about the importance of the different modes of transport in the country you have chosen. Use any sources available.

b Just-in-time is a principle whereby the suppliers of parts to large automobile manufacturers, for example, have to deliver their products exactly when they are needed for the production process. **List the advantages and disadvantages of this principle.**

Example: For: Less storage space is needed.
Against: An increasing number of lorries will block the roads and motorways.

c Find out about the latest developments in transport technology.

Grammar Overview

Present Tenses

Present simple: construction

I **like** / you **like** / he (she / it) **likes** we **like** / you **like** / they **like**	computer games.	Ajoutez **-s** (ou **-es**) à la 3e personne du singulier.

Present simple: uses

• Most cars **run** on petrol.	• Pour des vérités générales et généralités.
• I **get up** at 6. I usually **start** work at 7 o'clock. I never **eat** in the canteen.	• Pour des activités habituelles, notamment avec des adverbes de fréquence : **usually, always, never, sometimes, often...**
• He **doesn't like** shiftwork.	• Pour des habitudes et des préférences.
• My train **leaves** at 10 o'clock tomorrow morning.	• Pour des événements dans l'avenir qui sont planifiés à l'avance (horaires de train, etc.).

Present continuous/progressive (V + -ING): construction

I **am** he/she/it **is** we/you/they **are**	progress**ing** very well. mak**ing** a lot of noise. leav**ing** now.	Ajoutez **-ing** à la base verbale (V). Supprimez le **e** final après une consonne seule (**make**, **live**, **have**, **leave**, etc.).

Present continuous/progressive (V + -ING): uses

• I **am reading** a manual at the moment.	• Pour les choses qui se passent au moment où l'on parle…
• **I'm staying** at the Royal Excelsior Hotel.	• …ou qui ont une durée dans le présent.
• **I'm meeting** Ms Winter next week.	• Pour les activités (personnelles) planifiées.
• He's **always complaining** about his new boss.	• Pour critiquer une activité habituelle (verbe accompagné de **always**).

Past Tenses

Past Simple: construction

I work**ed** in London last year.	Ajoutez **-ed** ou **-d** à la base verbale des verbes réguliers.
He **met** Mr Miller yesterday.	Il faut apprendre les formes des verbes irréguliers. **Met** est le passé simple de **meet**.

Past Simple: uses

• I **finished** my studies	in 1998. 3 years ago. last week. yesterday. on 30th June.	• Pour des actions qui ont eu lieu dans une période de temps révolue. Normalement, on précise le moment ou la date : **yesterday, in (1990), two weeks ago, last year, last month, last week...**

Present Perfect

Present perfect simple (has/have + V + -EN): construction

I / you / we / they **have** He / she /it / **has**	work**ed** well.	Pour les verbes réguliers, utilisez **have** ou **has** + **-d** ou **-ed** (**lived, worked**...). Le participe passé est identique au passé simple.
I **have seen** her new car.		Il faut apprendre les participes passés des verbes irréguliers, ex : **see / saw / <u>seen</u>.**

Present perfect continuous/progressive (be + -EN + V + -ING)

I / you / we / they **have been** he / she / it **has been**	work**ing** for a year.	Construction : **have/has** + **been** + la base verbale + **-ing.**

Present Perfect; uses

• I **have written** 3 letters.	• Pour les actions terminées (souvent récentes) sans indiquer un moment précis.
• We **haven't had** any complaints **so far**. • **I've worked** in many places **in my life**.	• Avec des expressions comme **so far, this week, this year, in my life …** L'activité ou l'état a commencé dans le passé mais continue dans le présent.
• I **have had** the car **for** 5 years. • I **have known** the company **since** 1998.	• Avec **for** (durée : **2 hours, 3 weeks**, etc.) ou **since** (moments dans le passé : **2001, January, Christmas, last week**, etc.). L'activité a commencé dans le passé et continue dans le présent. Elle n'est pas terminée.
• I **have been living** in London **for** 20 years. • I **have been living** in London **since** 1982.	• La forme progressive, en **-ing**, met l'accent sur la continuité de l'action.

Comparison of adjectives

Construction

fast: fast**er**, **the** fast**est** easy: easi**er**, **the** easi**est** hot: hot**ter**, **the** hot**test** clever: clever**er**, **the** clever**est** narrow: narrow**er**, **the** narrow**est**	Ajoutez **-er** ou **the …est** aux adjectifs "courts" (d'une ou parfois de deux syllabes). Pour les adjectifs qui se terminent en **-y**, remplacez le **-y** par **-i**. Doublez la dernière consonne des adjectifs qui se terminent en **-t**, **-d**, **-g**, **-m**, **-n** ou **-p**.
nervous: **more** nervous, **the most** nervous splendid: **more** splendid, **the most** splendid expensive: **more** expensive, **the most** expensive beautiful: **more** beautiful, **the most** beautiful	Pour certains adjectifs de deux syllabes et tous les adjectifs plus longs, ajoutez **more** ou **the most**.
good: **better, the best** bad: **worse, the worst** much/many: **more, the most** little: **less, the least**	Notez les formes irrégulières de certains adjectifs.

Uses

• He is **as** clever **as** I am / **as** me. *(aussi … que)* • Mike is **less/more** helpful **than** Tim. *(plus/moins … que)*	• Comparison d'égalité : **as … as** • Utilisez **than** pour la comparaison de supériorité ou d'infériorité.
• The new electric drill is the most expensive tool **in** the brochure.	• Utiliser **in** dans des expressions comme : **in the world.**

Adverbs

Construction

She **is often** late. He **has never worked** in a factory before.	Placez les adverbes de fréquence avant le verbe, sauf quand ils accompagnent le verbe **be**.
clear**ly**, usual**ly**, quick**ly**, bad**ly**, easi**ly**, happi**ly**, tru**ly**	Pour changer un adjectif en adverbe, ajoutez **-ly** à l'adjectif. Le **-y** final de l'adjectif devient **-i**.
daily, weekly, monthly, yearly, early	Certains mots ont l'air d'être des adverbes mais ils sont à la fois adjectifs et adverbes.
He drives **fast**. She works **hard**. They arrived **late**.	Ces mots aussi sont à la fois adjectifs et adverbes.
The company pays **well**.	L'adverbe de **good** est **well**.
fast: fast**er**, fast**est** carefully: **more** carefully, **most** carefully	La comparaison des adverbes suit les mêmes principes que la comparaison des adjectifs.
well: **better, best** badly: **worse, worst**	Notez ces formes irrégulières.

Uses

• The new machine **works well**.	• Un adverbe qualifie un verbe…
• The small printer makes **remarkably clear** printouts.	• … ou un adjectif…
• Our new machine works **extremely quickly**.	• … ou un autre adverbe.

Word order: *questions fermées*

Construction

Are they designing a new power tool? **Have** you got the new brochure? **Have** you met Mr Miller? **Can** you supply spare parts?	Si les auxiliaires **be, have** ou un modal (**can, should**, etc.) sont visibles à l'affirmatif, ils sont placés en début de phrase, avant le sujet, pour former une question.
Do you/**Does** he smoke? **Did** the manager see you?	S'il n'y a ni auxiliaire ni modal apparent, utilisez **do/does/did** en début de question.
Are the visitors **having** lunch in the canteen? **Do** you **have** any idea of how to repair this?	Notez que **have** n'est pas uniquement un auxiliaire ; il peut aussi être un verbe.

Uses (*questions fermées* and short answers)

• "**Does** Sally speak French? " "Yes, she **does**. / No, she **doesn't**." • "**Can** these tools be used in milling as well as in turning? " "Yes, they **can**." / "No, they **can't**." • "**Is** the lathe set properly for this job?" "Yes, it **is**." / "No, it **isn't**." • "**Did** he meet the Japanese at the fair?" "Yes, he **did**." / "No, he **didn't**." • "**Will** you finish this job by Friday?" "Yes, I **will**. / "No, I **won't**."	• La réponse courte reprend le même auxiliaire (ou modal) qui figure dans la question. On l'appelle "réponse courte" parce que le verbe n'est pas repris dans la réponse.

Word order: *questions ouvertes*

Construction

How do you **go** to work? **What** does she do? **Where** do you live? **Why** doesn't the machine work? **When** is the delivery expected? **Which** tools do you need for the job? **Whose** are these gloves?	Lorsque le mot interrogatif (**How, what, where**, etc.) est le complément d'objet direct du verbe, inversez la phrase comme pour les questions fermées.
Who operates this machine? **What** causes the short circuit? **Whose** salesmen visited the fair?	Il n'y a pas d'inversion lorsque le mot interrogatif est le sujet du verbe.

Defining relative clauses
Uses

• The man **who/that** is operating the CNC machine works at the training centre.	• Pour les personnes, utilisez **who** ou **that**, si le relatif est le sujet du verbe qui suit.
• The machine element **which/that** is moving the tool is the main spindle.	• Pour les objets, utilisez **which** ou **that**, si le relatif est le sujet du verbe qui suit.
• Tom is an engineer **whose** skill is incredible.	• Pour le génitif (possession), utilisez **whose**.
• They called the technician (**who**) they saw standing at the milling machine. • Sally read the operation plan (**which/that**) she had filled in that morning.	• Lorsque le relatif est le complément d'objet du verbe qui suit, on peut omettre le relatif.

Non-defining relative clauses
Uses

The relative as subject: • Sally's friend**, who is a skilled worker,** is doing his final exams this summer. • The machine**, which is our latest model,** works perfectly. The relative as object: • The three French apprentices**, who Sally met at the training centre,** are in their last year of training. • The machines**, which/that Sally used at the training centre,** are now obsolete.	• Ces propositions sont placées entre virgules. Elles donnent des informations sur le nom qui les précèdent. Attention ! Les virgules sont importantes. Notez la différence entre ces deux phrases : **The workers who were late were sacked.** (Ceux qui étaient en retard furent licenciés ; d'autres n'étaient pas en retard et n'étaient pas licenciés.) **The workers, who were late, were sacked.** (Tous les ouvriers étaient en retard et furent licenciés.)

Reported speech
Construction and uses

• “We **are measuring** this component.” He **says** (that) they **are measuring** the component.	• Si le verbe introducteur est au présent (**says**) le temps ne change pas dans la subordonnée. **That** peut être omis.
• He **said** (that) they **were measuring** the component.	• Si le verbe introducteur est au passé, dans la subordonnée le présent devient le passé, …
• “The trainee **forgot** to set the speed.” He **said** (that) the trainee **had forgotten** to set the speed.	• … le passé (simple et progressif) et le **present perfect** deviennent le **past perfect**, …
• “The drilling **will** have to be prepared.” Oliver **said** that the drilling **would** have to be prepared.	• … **will** devient **would**, …
• “**I** can’t find the measuring equipment.” Sally **told me** (that) **she** couldn’t find the measuring equipment.	• … il faut changer certains pronoms (notez aussi qu’il faut utiliser **tell** lorsqu’on précise à qui on parle), …
• “We are seeing him **tomorrow**. She added that they were seeing him **the next day**. • “I programmed the machine **two days ago**.” He said that he had programmed the machine **two days earlier/before**.	• … et il faut changer les adverbes de temps.
• “**Use** HSS twist drills for drilling it.” **The instructor advised Sally to use** HSS twist drills for drilling it.	• Utilisez une construction avec l’infinitif pour transformer les ordres en discours indirect : **told … to, advised … to**, etc.
• “**Why has Paul used** the impact drill?” **He wanted to know why Paul had used** the impact dril. • “**When did you sharpen** the drill, Peter?” **He asked when Peter had sharpened** the drill. • “Sally, **why don’t you** call the salesman?” He asked Sally **why she didn’t** call the salesman.	• Notez la façon dont on transforme les questions ouvertes en questions indirectes.
• “John, **are** the new machines **working** OK?” He asked John **if/whether** the new machines **were working** OK.	• Notez la façon dont on transforme les questions fermées en questions indirectes. Utilisez **if** ou **whether**.

The passive voice

Construction

When the visitors **are taken** to the factory, make sure they **are provided** with protective clothing.	Pour construire le passif, utilisez **be** + le participe passé.

FORME ACTIVE	**We**	**recharge**	**the batteries.**
	sujet	*verbe*	*COD*
	sujet	*verbe*	*objet (**by** + agent)*
FORME PASSIVE	**The batteries**	**are recharged**	**(by us).**

Uses

• When the goods are ready, they **are transported** to the customer **by** lorry.	• On utilise très souvent le passif en anglais. Le français préfère souvent une construction avec "on". Il n'est pas toujours nécessaire de préciser l'agent (**by**…).
• The lathes **were shipped** to France yesterday. • The milling machines **have been shipped** to Germany. • The grinding machine **will be shipped** to London tomorrow. • The goods **would be sent** by plane if costs were lower. • Everything **should have been shipped** yesterday, but there was a delay.	• Quelques exemples du passif : présent, passé, futur, conditionnel…
• He has had an accident and **is** now **being taken** to hospital. • The order **was being prepared** when the customer rang to complain about the delay.	• On trouve souvent la forme progressive (V + -ING) du passif, surtout au présent et au passé.

The gerund
Construction

check + **ing** = checking drill + **ing** = drilling	Le gérondif est identique au participe présent (V + -ING)
put – put**ting** control – control**ling** travel – travel**ling** refuel – refuel**ling**	Il faut doubler la dernière consonne si elle est précédée d'une voyelle. Aux États-Unis, on ne double pas le dernier **-l** (**traveling**…).
file – fil**ing** measure – measu**ring**	Supprimez le **-e** final, mais notez : **agree – agreeing**, **see – seeing**

Uses

• **Filing** is rarely done by hand.	• Le gérondif s'emploie en tant que sujet d'un verbe, tout comme un nom.
• He was afraid **of making** a mistake. • They objected **to trying** the new software **without having** a training session first. • The manager apologised **for being** late. • They prevented the visitors **from entering** the production area without a hard hat.	• Les prépositions sont toujours suivies d'un gérondif ou d'un nom. (Notez que **to** peut être une préposition : **look forward to, get used to…**)
• You must **avoid overheating** the machine. • He **recommended turning** the machine off. • We cannot **risk losing** this customer. • They **suggested using** a different method. • He **enjoys working** in Italy.	• Un verbe suivi d'un infinitif en français peut se traduire par V + **to** + V (**want to do, hope to do…**) ou par V + gérondif. Les principaux verbes qui sont suivis d'un gérondif sont : **stop, start, continue, imagine, mind, enjoy, suggest, avoid, risk, deny, practise, keep, can't help, can't stand, postpone.** Notez aussi : **feel like, spend time…** + le gérondif.
• He **remembered seeing** the man before. • He **remembered to switch off** the machine. • She **stopped smoking** last year. • She **stopped to look** in a shop window.	• Ces verbes sont suivis de V ou d'un gérondif selon leur sens : **remember** + V + -ING = se rappeler… **remember to** + V = ne pas oublier de… **stop** + V + -ING = arrêter de… **stop to** + V = s'arrêter pour…
• They **tried oiling** the hinge but it still made a noise. • They **tried to open** it, but they couldn't.	• **Try** + V + -ING s'emploie lorsque le deuxième verbe est le moyen d'atteindre un objectif ; **try to** + V s'emploie lorsque le deuxième verbe est l'objectif lui-même.

Articles and expressions of quantity
Uses

• Turn off **the** power drill, please. • **The** coolant that you're using isn't the best on the market.	• L'article **the** s'emploie au singulier et au pluriel, devant les noms dénombrables ou indénombrables.
• I like **ø** apples. • **ø** Petrol is more expensive than **ø** diesel oil.	• L'article zero (= absence d'article, ou **ø**) s'emploie pour des généralités, devant les noms dénombrables ou indénombrables.
• There are **some** spare parts in the garage. • There is **some** copper wire over there. • We don't use **any** seasonal workers. • Is there **any** dirt in the machine? • There are **no** cars in the car park? • **No** employee has to wear a safety mask. • **A lot of** modern factories are robotised.	• **Some, any,** (quelques, des) **no** (aucun) et **a lot of** (beaucoup de) sont des expressions de quantité qui s'emploient avec des noms dénombrables ou indénombrables. On utilise **some** dans les phrases affirmatives, **any** dans des questions et phrases négatives.
• Would you like **some** coffee?	• Pour certaines questions (comme celle-ci), cependant, on utilise **some** parce que ce n'est pas une "vraie" question ; c'est une offre. Le locuteur sait déjà qu'il y a du café.
• **Anyone** can do this job. • I'll work **anywhere** if the pay is good.	• Notez le sens des composés de **any** dans les phrases affirmatives : n'importe qui, n'importe quand, etc.
• **A little** oil may be needed. • I noticed **a few** defects in the latest batch. • **Few** people work overtime here.	• **(A) little** ([un] peu de) s'emploie avec des indénombrables, **(a) few** (quelques) avec des dénombrables.
• There are**n't many** men in the workshop. • I have**n't** got **much** time, I'm afraid.	• Utilisez **much** (beaucoup de) avec des indénombrables, **many** avec des dénombrables. On préfère **a lot of** à l'affirmatif.
• **How many** (parts) do you need? • **How much** does it cost?	• Notez l'emploi de **how much** et **how many** (combien de).
• **All (of)** the employees **are** going on strike. • **The whole** factory **is** going on strike.	• Ne pas confondre **all** (qui s'emploie pour un groupe) et **the whole** (pour une unité).
• **Most** computers are sold with software. • **Most of** the circuits are miniaturised. • **Most of them** are sold with software. • **All of them** are going on strike.	• **Most** (la plupart [de]) s'emploie de deux façons, avec ou sans **of**. Comme pour **all**, il faut insérer **of** s'il est suivi d'un pronom.

Modal auxiliaries

Construction

The principal modals: **will, would, should, can, may, could, might, must**	Les auxiliaires modaux s'emploient pour exprimer un avis personnel : émettre des hypothèses, prononcer un jugement, exprimer une possibilité ou une certitude, donner la permission ou interdire quelque chose…
He **should** see a doctor. They **must** make a report.	Les modaux sont invariables, ni singulier ni pluriel. Ils sont suivis directement de la base verbale (V).
We **must** leave/We **have to** leave/We **have got to** leave. You **mustn't** smoke in here/You **aren't allowed to** smoke in here. He **can** do the job very quickly/He**'s able to** do the job very quickly.	On peut utiliser d'autres verbes pour remplacer certains modaux.
You **can** go home now. (present) You**'ll be able to** go home soon. (future) He **was able to** go home early yesterday. (past) She **must** be more careful. (present) She**'ll have to** be more careful. (future) She **had to** be more careful. (past)	Il faut utiliser ces autres verbes pour exprimer le passé ou le futur de **can** et de **must**. Notez que **could** exprime une compétence (et non une possibilité) dans le passé, ex : **he could run very fast when he was younger.**

Uses

He isn't at the office, so… • he **must be** ill. • he **must be** working at home. • he **must have** had an accident.	Des modaux: • pour exprimer la certitude
• he **may/could/might be** on holiday. • he **may/could/might be** taking the day off. • he **may/could/might have** overslept.	• pour exprimer la probabilité
• he **can't** attend the meeting. • he **can't have** seen Mr Miller.	• pour exprimer l'impossibilité
• You **can/may** see Mr Miller now.	• pour exprimer la permission
• You **must** read the instructions carefully. • You **can't** take your lunch break now. • You **mustn't** take your mask off.	• pour exprimer l'obligation et l'interdiction

• He **can** speak English quite well but he **can't** speak Spanish. • Everyone **could** understand the instructions.	• pour exprimer la compétence/savoir-faire
• You **should** be more careful. • You **shouldn't** work so hard.	• pour donner des conseils
• **Could/Would/Will/Can** you open the window, please?	• pour demander quelque chose poliment
• You **need to** go on a training course. • That machine **needs** lubricat**ing/needs to** be lubricated. • It didn't **need** repair**ing**. You **needn't** worry.	• **Need** est un cas à part. On l'utilise soit comme un verbe "normal" soit comme un modal suivi du gérondif (V + -ING).
• You **don't have to** wear safety boots.	• Il ne faut pas utiliser **mustn't** (tu ne dois pas…) pour une absence d'obligation. Utilisez **not have to** (tu n'est pas obligé de…).

Conditional sentences (If…)

Construction and uses

• **If** you **lubricate** this old machine regularly, it **will run** smoothly for a long time.	• Type I (probable) **If** + présent; **will** + V (ou **is going to**)
• **If** you **lubricated** this old machine regularly, it **would run** smoothly for a long time.	• Type II (peu probable) **if** + passé simple, **would** + V
• **If** you **had lubricated** this old machine regularly, it **would have run** smoothly for a long time.	• Type III (impossible) **if** + *past perfect* (**had** + V -EN); **would have** + V -EN
• **If** you **heat** ice it **melts**.	• Pour les lois de la nature, etc., utilisez le présent dans la principale et la subordonnée.
• **If I were you,** I wouldn't use a combination spanner here.	• Pour donner des conseils, utilisez cette construction (**were** au lieu de **was**).
• **If only** I **earned** more money! • **If only** we **had listened** to the instructor!	• On peut utiliser **if only** au lieu de **I wish**.

Expressing the future
Construction and uses

• I think most of the team **will** take their holidays in August this year.	• **Wil**l + V s'emploie pour les prédictions ou pour le déroulement prévisible des choses.
• **I'll** do my best to get the job. • Wait a minute, **I'll** help you	• **Will** + V s'emploie aussi pour des promesses et des décisions spontanées.
• I'**m going to** apply for a job in teleworking.	• **Going to** + V exprime une intention de faire quelque chose.
• Look at those clouds! It**'s going to** rain.	• **Going to** + V exprime également la détermination : l'observation permet de prédire un événement.
• Jim's train **arrives** at 4.30 p.m. • Our office **opens** at 9 o'clock tomorrow morning.	• Le présent simple s'emploie pour des événements planifiés bien à l'avance (horaires de train, d'avion, etc.).
• Jennifer **is meeting** her colleagues this evening.	• Le présent progressif (**be** + V + -ING) s'emploie également pour des événements planifiés, mais au niveau personnel (agenda, etc.).

UNIT 1

B Exercise a

Bill: Good afternoon, Michel. How do you do?
Michel: How do you do?
Bill: Did you have a pleasant flight?
Michel : Yes, thank you. Although there was a bit of a delay.
Bill: Sorry to hear that. Please take a seat. Would you like something to drink? A cup of coffee or perhaps some tea?
Michel : Tea, would be fine, thank you.
Bill: Milk and sugar?
Michel : Just milk, please. Thank you.
Bill: Right. Let me introduce you to your colleagues. This is Roberto Garcia from Spain ...
Michel : Pleased to meet you.
Roberto: Pleased to meet you, too.
Bill: ... and Jerry Miller from the USA.
Jerry: Nice to meet you, Michel.
Michel : Nice to meet you.
Bill: ... and this is Annette Mair from Austria. I believe you've met before?
Michel: Yes, we met at the interview. Nice to see you again, Annette.

After some small talk Bill Walker introduces Steven Hill.

Bill: Ah, there's Steven. Michel, may I introduce you to your new boss? This is Steven Hill, who is in charge of production, engineering and quality control.
Steven: How do you do? Pleased to meet you and welcome to Power Engines. I hope you'll enjoy working for us.
Michel: I'm sure I will.
Bill: Before Steven shows you around the various workshops, I'd like to give you a brief outline of our company ...

C Exercise c

1. Michel Arnaud drives in at the main entrance to the premises of Power Engines Ltd., but he finds a parking space only in the lot near gate 2. He leaves the lot and walks left, past the administration building and the canteen, and directly ahead of him is the octagonal visitors' centre. He signs in there and follows the signs to the building attached to it, the (?), where he plans to look around and get an idea of the history of the company before meeting the personnel manager, Bill Walker, for a tour of the premises.

2. The tour of the company begins in the research and development (R&D) centre, right next to the museum. After John Smiley has described the projects that are being planned, Michel is taken next door to the laboratories, where his future colleagues Roberto, Jerry and Annette show him what they are working on. When they leave the labs they turn left, and left again, and enter the (?).

3. They walk out of the production area and turn right. They walk past the washrooms on the left and leave the building. They turn right and pass the administration building on their left. They finally arrive at the (?), opposite R&D.

C Exercise e

Good afternoon, ladies and gentlemen. Before we take you on a tour of Power Engines Ltd, I would like to explain the itinerary and the set-up of our plant. You are now at the Visitors' Centre. From here we'll go first of all to the administration building, where Emily Miller will give you a brief overview of Power Engines: who we are, what we produce and who our customers are. Emily will take you to the research centre across the yard, where the head of our Research & Development Department, John Smiley, will tell you something about fuel cell technology. His assistant, Peter Thompson, will take you to the

aboratories where, hopefully, you'll be able to watch an ongoing experiment.
Steven Hill, who is responsible for Production and Engineering, will then take over and show you some interesting aspects of our production unit. Jane Owttrim, from Purchasing and Stores, will explain how we organise our supplies and after that our public relations officer will give you some historical data and outline some of our plans for the future.
Finally you'll have the opportunity to relax in our canteen where Bill Walker, the personnel manager, will describe our strategies to recruit scientists and other experts in our field from all over Europe.
We wish you a pleasant and informative visit to Power Engines Ltd., and thank you for your attention.

UNIT 2

B Exercises a and b

Michel: Power Engines, Michel Arnaud speaking.
Mr Slate: Hello, this is Andrew Slate from Distel Electronics. Could I speak to Steven Hill?
Michel: Yes, certainly. Hold on, I'll put you through. – I'm afraid there's no answer; he must be in a meeting. Can I take a message, Mr Slate?
Mr Slate: Er, yes please. We made you an offer for the supply of control panels. Er, we've found some new suppliers for our components. They can deliver the same quality at a better price. So we are now in a position to make you a more attractive offer: we can deliver the same panels at a 15% reduction in price. Terms of delivery and payment will be the same, but we would need your answer soon, let's say by Tuesday morning, Tuesday the 12th, that is. Have you got that? I'm sorry to put you through all this.
Michel: No problem at all, sir. I'll just read that back. You are making a new offer, reducing the price by 15%, and you need our answer by Tuesday morning.
Mr Slate: Perfect, thanks a lot. Have a nice weekend.
Michel: Thank you.
Mr Slate: Oh, before I forget, in case I'm not in on Tuesday, Mr Hill should ask for Ms Auburn.
Michel: I'm sorry, could you spell the name please?
Mr Slate: AUBURN – A for Andrew, u for uncle, b for Benjamin, u for uncle again, r for Robert, and n for Nellie. Her extension is 274.
Michel: Thank you, I've got it. Have a nice weekend, yourself, Mr Slate.

B Exercise f

1.
Woman: Directory enquiries, can I help you?
Caller: I'd like a Portsmouth number, please.
Woman : Yes, and the name, please?
Caller: Jackson and Green Ltd.
Woman : The area code is 01705 and the number is 9133407.

2.
Woman : Directory enquiries, can I help you?
Caller: I need a number in Sweden, please.
Woman : Yes, which town, please?
Caller: Falun, that's F-A-L-U-N. Sörensen Software in Falun.
Woman : The country code for Sweden is 0046, then you dial 23 – that's the area code – and the number 4978.

3.
Woman : Directory enquiries, can I help you?
Caller: Yes, please. I'm trying to find a German number. It's MPX Maschinenteile in Eschwege, Germany.
Woman : Right. Er, the country code for Germany is 0049, the area code is 5651 and the number is 567481.

4.
Woman : Directory enquiries, can I help you?
Caller: I'd like a fax number, please. It's Charles Brockden Brown Consultants in Philadelphia, Pennsylvania, USA.

Woman : One moment, please. Hello?
Caller: Yes?
Woman : The country code is 001, the area code is 215, the telephone number of the company is 3451985 and their fax number is 3451930.

B Exercise i

1. http://www.ft.com (http colon double slash www dot ft dot com)
2. http://www.irish-times.ie (http colon double slash www dot irish hyphen times dot ie)
3. http://www.city.net/countries/united_kingdom/england/london (http colon double slash www dot. city dot net slash countries slash united underline kingdom slash england slash london)
4. http://www.virtual-london.co.uk/ghosties.html (http colon double slash www dot virtual hyphen london dot co dot uk slash ghosties dot html)
5. Site.Feedback@ft.com (Site dot Feedback at ft dot com)
6. info@phantasy.de (info at phantasy dot de)

B Exercise k

1. 1 pint
2. 10 inches
3. 3 quarts
4. 5 gallons
5. 2 ounces
6. 50 yards
7. 2 acres
8. 100 miles

UNIT 3

A3 Exercises a and b

Denise: Good evening, I'd like to update my computer. Could you tell me a bit about the latest hardware and software that's available?
Asst: Certainly. Can I ask what kind of computer you've got?
Denise: It's a 433-MHz computer, the hard disk has got 8 GB, it has 64 MB memory, and, well, what else have I got? Oh, yes, a CD-ROM drive.
Asst : Hm, that's pretty old stuff, you know. I'm not sure whether you can really update it. What will you need your computer for?
Denise: Well, first of all I'll need it for CAD, then of course I want to be able to use a word processor and a spread-sheet program.
Asst : Your computer will be OK for word processing and spreadsheets, but the new CAD programs need a lot more memory and a faster CPU. What about the Internet?
Denise: Yeah, I think I'll need access to the Internet – we are supposed to get some practice in creating websites.
Asst : You see, it's very important to have a fast computer; the faster it is, the faster you get your information from the web. What about a scanner?
Denise: The longer I think about it, the more I think a scanner would be quite useful, but I'm afraid I won't be able to afford it right away.
Asst : That's all right, you can buy one at a later date. They've gone down a lot in price lately and prices are still falling.
Denise: While we're at it, would there be any chance of hooking up a digital camera, too?
Asst : You'd need a lot of memory for that and you'd need a DVD drive for storage, too. I can't see how we can fit all that into your old computer. Let's see. I'll tell you what I'll do. I'll write down what kind of hardware you need and I'll work out a good price for you.

A5 Exercise a

Help desk: Hello, this is help desk 17 at Eastshore Entertainment; my name is Linda Euston. Can I help you?
Customer: I've been trying to reach you for over ten minutes! How long does it take to pick up a phone?
Help desk: I'm really sorry, but we are very busy

at the moment. How can I help you, sir?

Customer: I've got this game called Invade and Take over. I'm having problems with the 3D sound. The game doesn't accept the default settings of my 3D sound card.

Help desk: So, you have no sound at all?

Customer: All I can hear is a screeching sound.

Help desk: I see. I can understand that this is very annoying. Which version of Invade and Take over do you have, sir?

Customer: I think it is the 3.0, second release. Do you need my serial number?

Help desk: No, thank you, that won't be necessary. What I do need to know is what kind of sound card you have.

Customer: I have a 3D sound card from a company called Creative Tech.

Help desk: Oh yes, we've had several problems with this sound card since the release. The problem's quite easy to fix. Do you happen to have access to the Internet?

Customer: Yes, I do.

Help desk: This is what you should do: Go to our Website; the address is www.East shore.com – the address is also on the last page of your manual. Click under downloads and select patchCT for Creative Tech.

Customer: Does it take very long to download?

Help desk: That depends on your connection, but it shouldn't be any longer than 10 minutes.

Customer: And you can guarantee that it works?

Help desk: We've never had any problems so far. If it doesn't work, you can call us any time on 1 800 555 267. Anyway, please give me your address; we would like to send you the Invade and Take over mousepad and T-shirt free of charge to make up for the problems you've experienced.

Customer: Oh, that's very kind of you, and thank you for your help.

Help desk: You're welcome. It was a pleasure to serve you, and please don't hesitate to call again if there should be any other problems.

Customer: Thank you, I will. Oh, before I forget, my address is ...

UNIT 4

A1 Exercise a

Christof: This is our general workshop, where our apprentices are trained in mechanical engineering. In the centre are the work benches with a vice on each bench plate.

Bob: And between the large windows over there I can see two column drills, and there is also a centre lathe next to that blue door, isn't there, and a milling machine opposite the lathe?

Christof: Yes, that's right. Some of our training units include basic machine training, too, so our young apprentices get their first experience of the machinery they'll be operating later as skilled workers.

Bob: I see. Now, could I have a look into your CNC shop, Christof? It's the heart of your manufacturing centre, isn't it?

Christof: Yes, you're right. It's just behind that blue door. That's where you'll see our latest lathe model with a high-speed spindle doing up to 10,000 revolutions per minute. Do you remember? We demonstrated it at the Hanover fair?

Bob: Oh, yes, and your fantastic milling machine ...

Christof: ... with the option of 30-tool-capacity tool changers.

Bob: Exactly! Oh dear, I've got some machine oil on my hands from one of the vices. Do you think I could wash them somewhere before we go on?

Christof: Certainly, the washrooms are next to the entrance of this workshop. Just go back into the corridor and you'll see them on the right. The men's washroom is the first door. Well, the signs will help you anyway.

B5 Exercise e

Customer: I'm interested in buying a driver drill. What would you recommend?

Assistant: Well, let me show you our latest model. It's a reversible cordless driver drill with a totally new design. It comes with a twospeed variable trigger switch, and you can set five torque adjustments for superior power.

Customer: Sounds impressive. What about the service life of the battery?

Assistant: This model comes with two batteries and their service life is more than 2 hours of operation. An interesting new feature is the red signal lamp that indicates low charge. For recharging there is this quick charger.

Customer: What do I do with empty batteries that can't be recharged any more? You know, I can't dispose of them with the domestic waste.

Assistant: That's no problem. We are happy to take those batteries back and dispose of them in an ecologically friendly way.

Customer: Well, you said something about quick recharging.

Assistant: Yes, it's a 9.6-V DC 1-hour rechargeable battery cartridge and a 1-hour charger with an auto shut-off.

Customer: Great. What about safety?

Assistant: That's another improvement. The drill has a new safety feature that locks out the power when not in use.

Customer: Fine. Does the carrying case have enough space for drill bits?

Assistant: Yes, of course. It has separate compartments for bits, battery cartridge and charger. And the material the case is made of is 100% recyclable.

UNIT 5

A

Dave: What's the problem, Steve? Any trouble with the machine?

Steve: Yes, it's not working and I don't know why.

Dave: Did you check the main switch?

Steve: Yes, of course I did. That was the first check and I checked the safety button as well.

Dave: Well, let's see. Look here, Steve, the safety button is still locked in, you see? To release it, you must turn it a bit to the right and then pull it out, okay?

Steve: All right. And could you please check if the machine is set correctly?

Dave: Hmm, the spindle's set fine and the drill's properly clamped in the chuck. But what about the revolution speed ?

Steve: Well, as the drill is an HSS centre drill I set the speed as high as possible which is 25,000 revolutions per minute.

Dave: Fine, and you know that after centring you read the required revolution speed from the table here at the front of the machine gear box?

Steve: Yes, I know. The diagram in the table tells you the ratio of revolution speed and diameter of the drill.

Dave: And think of the machine table. It should be set high enough for you to drill properly into the material.

Steve: Oh yes, you're right, thank you. And the machine vice must be in central position as well?

Dave: Yes, you've got it. If you have any more problems, just give me a ring…

Steve: Will do. And thanks again!

C2 Exercises b and c

Woman: Star Tools, service station. My name is Susan Brown. Can I help you?

Customer: Hello, this is Bill Morris from DEP-Design and Engineering Performance. We've got a problem with your power drill, type Star 600 SP. The reversing lever doesn't work any more.

Woman : When did you buy the tool?

Customer: Er, about four months ago.

Woman : You'll be pleased to know that we

offer a six-month warranty on this particular product, in which case you are still within the warranty period.

Customer: Uh huh, what does that mean exactly?

Woman : Well, you can take the drill, along with your warranty or proof of purchase, to one of our approved service centres and they will either repair it or replace it. You'll find your nearest service centre on the back page of your manual. There's a list of addresses there.

Customer: Ah, that sounds pretty straightforward. Well, thank you very much for your help.

Woman : That's all right, sir, you're welcome. Goodbye.

UNIT 6

Starter Exercise b

Jennifer: Have you ever thought about the enormous amount of energy we consume in our private homes every year? It's really amazing, when you add up the consumption of our most common household equipment. There was a report on TV yesterday with some really interesting information. You wouldn't believe it!

Morris: Yes, I know – I'm sure we could save a lot more energy – it's the little things that help.

Jennifer: Yes, you're right. Let me give you a few of the examples I noted down. Did you know that the refrigerator is one of the most expensive household appliances to run? Its consumption of electricity amounts to 642 kW a year – closely followed by the TV set which is another 292 kW, and that's if it's only on for four hours a day!

Morris: Well, that's a lot, but it's nothing compared to our water heater, which eats up 2190 kW! Another 'little helper' is the dishwasher, that takes 432 kW. You see, I'm well informed. Guess why? – Well, I watched the programme, too.

Jennifer: I was surprised by the relatively low energy consumption of the microwave. I didn't expect that it's only 89kW. And most surprising is the vacuum cleaner. I expected a lot more than 38 kW.

Morris: At least the mobile phone is cheap as far as energy consumption is concerned. I think we can afford 20 kW per year in future. That's nothing compared to our telephone bill!

C2 Exercise a

John: Well, before we have a detailed look at the principle of the fuel cell, I'd like to give you some general background information on engine technology. As you know, the fuel cell generates power in a fundamentally different way compared to internal combustion engines – ICE for short – and storage batteries. ICEs burn fuel to create heat. This heat is then converted into mechanical energy and finally into motive power. The efficiency of this conversion process is greatly affected by losses of wasted heat and friction. In contrast, fuel cells convert fuel directly into electricity, making this new technology two or three times more efficient than internal combustion engines. Like an ICE, fuel cells use fuel from a tank that can be quickly refuelled. Unlike ICEs, however, fuel cells do not burn fuel and therefore do not produce air pollutants, the emissions resulting from combustion.

Michel: In other words the fuel cell is the answer to all the environmental problems we've had so far. If all engines change over to fuel cell technology within the next ten years, we'll solve one of our most important environmental problems. Is that right?

John: Exactly.

Michel: But what about electric cars running on batteries? Don't they have the same advantage as cars which are equipped with the latest fuel cell technology?

John: Well, that's not quite right. There are major differences in both technologies. Batteries do not create power. They are energy storage devices. As you know, batteries have to be recharged after a certain time. Batteries and fuel cells are both electrochemical devices that are known for their high efficiency, quiet operation and freedom from any polluting by-products of combustion. However, as batteries don't produce power, but only store it, they need central power plants to create the electricity which is needed to recharge them.

Michel: So the fuel cell creates its own power by changing hydrogen directly into electricity which is used for running a car, while a battery-driven car only uses up the stored power without replacing it.

John: Right. That's exactly the advantage of fuel-cell technology.

UNIT 7

A1 Exercises a and b

Allan: I think furniture should be made of real wood. That's a classic material. It's easy to work with, it's very strong and sturdy, it lasts a long time and you can repair it to a certain extent. And finally, if you want to get rid of your piece of furniture, you can always burn the wood ...

Heather: ... that is, if you have a wood stove and if the wood hasn't been treated or lacquered. There are environmental laws against burning wood. But to come back to your suggestion: People have made furniture out of wood for hundreds of years. I want to design something new. I think chairs and tables made of steel look quite elegant, especially if they are chromium-plated. Steel is much more durable than wood; you can form it into every shape you can think of. For tables, for example, you can have a steel construction and a glass plate. Or comfortable steel chairs with leather upholstery. Imagine how that would look!

Danielle: Come on, Heather, steel has been used for ages. My parents' furniture is made of steel frames. I hate those shiny chairs and tables. Let's use aluminium instead.

Heather: Aluminium's too soft, isn't it?

Danielle: Yes, I know. I wasn't talking about pure aluminium; I meant aluminium alloys.

Allan: Actually, I saw an ad for garden furniture the other day, and I believe it said it was made of aluminium. It is a very light material and resistant to corrosion. The problem is, you can't really paint it.

Danielle: I don't want to paint it, just an aluminium frame, an armrest made of black glass and leather-covered upholstery in a light grey.

Allan: OK, let's forget about wood, aluminium sounds good; it's something different anyway. What about you, Heather?

Heather: Maybe you're right. Steel is out, let's draft a few ideas and then ...

A1 Exercise f

The table I am describing is a side table. Its glass plate is an oval shape. It is about 3 feet long and 2 feet wide. It rests on a pyramid-shaped base about 15 inches high, the broad side on the floor, the tip supporting the glass plate. You sit on wooden cylinders, 1 foot 5 inches high and 16 inches in diameter. They are covered with leather. Next to your chairs you have big plastic 15-inch cubes on which you can put your cup or glass, magazines or a small lamp.

B5 Exercise e

1. polyamide	2. polyethylene
3. polystyrene	4. polyvinylchloride (PVC)
5. Teflon	6. acrylics
7. polymer	8. polyester

UNIT 8

A2

In the next 50 years the mean temperature of the earth could rise by between one and four degrees centigrade. A rise of four degrees centigrade is about the difference between the global temperature now and the temperature of the last ice age. This increase in temperature would first of all alter the world's climate and weather patterns. The tropics would get a little warmer and the polar regions could become dramatically warmer. In the past 20,000 years life on earth was able to adjust to this change. In future we will not have another 20,000 years to adjust. The most pessimistic calculations predict that we have less than 50 years to do so.

Water expands with heat. In the past century the world has become about half a degree warmer. Sea levels have risen by 10 to 15 centimetres. Many scientists now think that we should prepare for a 1-metre rise in sea levels in the course of the next five decades. Some say it could be two metres if the polar icecaps melt fast enough. All this is bad news for people who live at sea level. Droughts could become more frequent; so could storms. Agriculture could be disastrously hit; so could the forests. Ocean currents could alter, with Europe becoming colder as the Gulf Stream shifts.

UNIT 9

A1 Exercises d and e

Valerie: So, the two of us are going to fly to Philadelphia this August. How do we go about it, Chris?

Chris: Well, first of all I need the exact dates, and I suppose the best thing is to make a list of what we have to do. Let's see, we have another two months to go. Plenty of time to look for hotels, flights, car rentals.

Valerie: Smith Consulting will help us with that for sure. Let me see, the Tech Expo starts on August 29; I would like to arrive at least two or three days early, let's say on the 25th or 26th, so we won't fall asleep at our first business dinner. I also would like to talk to the people from Smith at least one day prior to the job fair.

Chris: As to the flights, I would really like to have a direct flight to Philadelphia, no stopover in New York or Boston. What about a hotel? I guess there are plenty. Where should we stay?

Valerie: We could always get one near the airport, but I'd prefer one near the city centre.

Chris: OK. So we need to check the flights and we must find a convenient hotel. Do we need a car?

Valerie: Maybe not for the first few days, since the job fair is pretty much in the centre and we can always take a cab, but I'd like to see something of the area too. Yes, a car for one or two days would be fine. Do you think we need an international driver's licence?

Chris: Yes, definitely. So car rental is settled. Now, do you think we could do something cultural too? After all, this city has a very attractive historic area, they have museums, for example the Rodin museum, theatres, orchestras ...

Valerie: That's fine by me. Let's find out what's on offer; after all, we want to know where to take our customers when everything is settled. And I hope we'll have time for some nice meals in the good restaurants; I love seafood.

B1 Exercise c

1. Attention please, paging Mr Fernandez, paging Mr Fernandez, please proceed to gate A 27.
2. Attention please, due to heavy fog all national flights will be delayed by at least one hour.
3. Attention please, all passengers flying to Brussels with a stopover in Lisbon, your flight has been cancelled. Please contact your airline information desk.
4. Attention please, all passengers arriving with TAP 434 from Lisbon, please proceed to the baggage claim area.
5. Attention please, will passenger Mary Watson, Mary Watson, please go to the information desk in terminal B to take a telephone call.
6. Attention please, this is the last call for passengers boarding TAP flight 433 to Lisbon.

UNIT 10

B1 Exercise a

Woman: Do I have the right of residence in the United Kingdom as a national of the European Union?

Official: European Community law grants nationals of the European Community the right to live and work in the United Kingdom.

Woman : What rights do I have if I want to work in the United Kingdom?

Official: You have the right to accept offers of employment or to be self-employed – for example, you can set up your own business. You don't need a work permit.

Woman : Do I need to apply for a residence permit or register with the police?

Official: As a European national it is not necessary. You are allowed to stay as long as you wish without obtaining a residence permit. There is no need to register with the police.

Woman : Can I live in the United Kingdom if I'm not working?

Official: If you have sufficient funds to support yourself throughout your stay without help from the British social security system, there is no problem. You have the same rights as in your home country.

Woman : Can I work in the United Kingdom if I'm studying?

Official: No problem. You can work in the United Kingdom during and after your studies.

Woman : What about my family?

Official: The members of your family may join you at any time. During your studies your family members are restricted to your spouse and your dependent children only.

Woman : Can the members of my family work in the United Kingdom?

Official: Certainly. European law applies to them in the same way as to you. They do not need work permits.

Woman : Do the same laws apply to all the other countries of the European Union as well?

Official: As this is common European law you can live and work in all the countries of the European Union.

Woman : Thank you for your help. It's good to know that I have the same rights in the United Kingdom as in my home country.

Official: Well, that is part of our common European idea.

Unit vocabulary

US	= American English	**pl.**	= plural
UK	= British English	**sing.**	= singular
v.	= verb		
n.	= noun		

UNIT 1 • Starter

anticipate *(v.)*	anticiper
applicant	candidat
application	demande d'emploi
appropriate	approprié, qui convient
available	disponible
challenge	challenge, défi
challenging environment	environnement stimulant
conduct *(v.)*	conduire
convenience	confort, commodité
correspondence	correspondance
cost accountant	comptable spécialisé en comptabilité analytique
currently	actuellement
CV (Curriculum Vitae)	CV
due to	à cause de
electrical engineer	ingénieur électricien
employment opportunity	perspective d'emploi
engine	moteur
engineering degree	diplôme / licence en ingénierie
ensure *(v.)*	assurer
essential	essentiel
fringe benefit(s)	avantages en nature
fuel cell	pile à combustible
human resources adviser	conseil en ressources humaines
initial research	recherche initiale
involve *(v.)*	comprendre, consister en
job vacancy	poste à pourvoir
maintenance	maintenance
manufacturing	fabrication
manufacturing engineer	ingénieur de la fabrication
mechanical engineering	mécanique
meet (needs) *(v.)*	répondre à (des besoins)
network support technician	technicien de maintenance de réseau
on file	sur fichier
particular specialization	spécialisation particulière
preferably	de préférence
quality engineer	ingénieur de contrôle de qualité
quote *(v.)*	annoncer, indiquer, citer
recruitment needs	besoins de personnel
requirement	exigence
research engineer	ingénieur chercheur
responsibility	responsabilité
skilled	qualifié
skilled worker	ouvrier qualifié
submit an application	poser sa candidature
thrive *(v.)*	réussir, se développer
varied	varié
wide area network (WAN)	WAN

A

baggage reclaim	livraison des bagages
customs	douane
departure	départ
newsagent	marchand de journaux
taxi rank	station de taxis

B

apologise *(v.)*	s'excuser, regretter
brief	bref
brief outline	idée générale
colleague	collègue
delay	retard
develop *(v.)*	développer
divorced	divorcé
drop off *(v.)*	laisser, déposer
engine test facility	installation pour tester les moteurs
hydrogen	hydrogène
in charge of	responsable de
marketing manager	responsable du marketing
public relations manager	responsable des relations publiques
purchasing department	service achats
recent	récent
self-employed	indépendant, qui travaille à son compte
self-employed person	personne qui travaille à son compte
shop	(ici) atelier

C

accounting	comptabilité
administration building	bâtiment administratif
advertisement / ad	annonce, (une) publicité
advertising	la publicité
attached to	attaché à
car park *(UK)*	parking
check the account	vérifier le compte
employee	employé, salarié
enquiry	demande de renseignements
feeder road	route d'accès
finished goods	produits finis
gate	porte, portail
letter of complaint	lettre de réclamation
main entrance	entrée principale
managing director	directeur général
market research	étude de marché
octagonal	octagonal
organigram	organigramme
parking lot *(US)*	parking
PR (public relations)	relations publiques
research and development	recherche et développement
storage	stockage
stores	provisions, réserves, stocks
survey	sondage

D Grammar revision

experience	expérience
graduate *(n., v.)*	diplômé ; obtenir son diplôme / sa licence
(Honours) degree	licence
(be) promoted	être promu, avoir une promotion
secondary school	lycée
short-term contract	contrat à durée déterminée

UNIT 2 • Starter

biro	stylo à bille, bic
envelope	enveloppe
hole punch	perforateur
notepad	cahier
paper clip	trombone
pencil sharpener	taille-crayon
stapler	agrafeuse
tray	plateau, casier corbeille

A

account	compte
area code	code région
assistant	assistant, adjoint
(get/hear the) busy signal	sonner occupé
button	bouton, touche
Can you manage alone?	Pouvez-vous vous débrouiller tout seul ?
certainly	certainement
complaint	réclamation
consist of *(v.)*	consister en
contract	contrat
control panel	tableau de bord
controls	contrôles
customer	client
dial *(v.)*	composer (un numéro)
do someone a favour *(v.)*	rendre un service à quelqu'un
do the filing *(v.)*	faire le classement
down the hall	à l'autre bout du couloir
either ... or	ou ... ou
engaged	occupé
enquiry	demande de renseignements
especially	particulièrement, spécialement
explain *(v.)*	expliquer
face down	face contre la vitre (photocopies)
(be) familiar with	bien connaître
files *(pl.)*	dossiers
format	format
I know my way round now.	Je connais bien les locaux maintenant.
I wonder if ...	Je me demande si ...
I'll be right with you.	J'arrive.
invoice	facture
occasionally	de temps en temps
of course	bien sûr
offer *(n., v.)*	offre ; offrir
order *(n., v.)*	commande, ordre ; commander, ordonner
overtime	heures supplémentaires
pass *(v.)*	passer
pigeonhole	casier (à courrier)
pile	pile, tas
polite	poli, bien élevé
provide *(v.)*	fournir
purchase *(v.)*	acheter
purchasing department	service achats
rarely	rarement
repeat *(v.)*	répéter
slot	fente
suggestion	suggestion, proposition
supplier	fournisseur

sure thing	bien sûr, naturellement
urgent	urgent
usually	d'habitude
Would you mind copying the faxes for me?	Pourriez-vous me copier les télécopies, s'il vous plaît ? (= Est-ce que cela vous dérangerait de … ?)

B

1 ounce equals …	1 once égale …
abbreviation	abréviation
agree on *(v.)*	convenir de, se mettre d'accord sur
altitude	altitude
answering machine	répondeur
Can I take a message?	Puis-je prendre un message ?
country code	code pays
deliver *(v.)*	livrer
diameter	diamètre
directory enquiries	renseignements
equal *(v.)*	égaler
equivalent	équivalent
extension	poste
extension number	numéro de poste
figure out *(v.)*	arriver à comprendre
hang on *(informal)*	ne quittez pas
hold on *(formal)*	ne quittez pas
I didn't catch that.	Je n'ai pas bien compris.
imperial measure (imperial system of measurement)	mesure britannique
lathe	tour
measure	mesure
message	message
period *(US)*	point
possible	possible
prepared	préparé / prêt
prevent *(v.)*	empêcher
pronounce *(v.)*	prononcer
reservoir	réservoir
roughly	approximativement
speak up *(v.)*	parler franchement, parler plus fort
speed limit	limitation de vitesse
spell *(v.)*	épeler
subject	sujet
telephone directory	annuaire
town hall	mairie
troy weight	troy
weight	poids
yellow pages	pages jaunes

C

abbreviate *(v.)*	abréger
abrasion	abrasion
addressee	destinataire
advertisement	annonce, (une) publicité
application	demande d'emploi, lettre de candidature
appreciate *(v.)*	apprécier
approval	approbation
approximately	approximativement
(for the) attention of	à l'attention de
bulk buyers	acheteurs en gros
business relation	relation d'affaires
by return / by return mail	par retour de courrier
chamber of commerce	chambre de commerce
charges	charges
complimentary close	salutation
current rates	taux en vigueur
distance	distance
enclosure	pièce jointe
equipment	matériel
favourable	favorable
introduce *(v.)*	présenter, introduire
lubricating oil	huile de graissage
manufacturer	producteur, fabricant
neat	net, soigné, bien rangé
(be a) nuisance	être énervant
paragraph	paragraphe
product range	gamme de produits
quote *(v.)*	annoncer, indiquer, citer
reduction	réduction
reference	référence
represent *(v.)*	représenter
respective	respectif
revive *(v.)*	raviver
rpm (revolutions per minute)	tr/min (tours à la minute)
salutation	salutation
signature	signature
subject line	sujet, référence
take into consideration *(v.)*	prendre en considération
terms of delivery	conditions de livraison
terms of payment	conditions de paiement
therefore	donc
tremendous	énorme
Yours faithfully	salutations distinguées

UNIT 3 • Starter

central processing unit (CPU)	unité centrale de traitement (UCT)
cursor	curseur

facilitate *(v.)*	faciliter
hard disk	disque dur
input device	périphérique d'entrée
joystick	manette de jeu
keyboard	clavier
laptop	portable
link *(v.)*	lier, connecter, chaîner
manipulate *(v.)*	manipuler
memory	mémoire
output device	périphérique de sortie
peripherals	périphériques
plotter	table traçante
provide *(v.)*	fournir
screen	écran
storage device	périphérique de stockage

A

access	accès
affordable	(dont le prix est) abordable
annoy *(v.)*	embêter
apologise *(v.)*	s'excuser, regretter
arrow	flèche
assistance	aide
backspace	(touche) retour
backup	(de) sauvegarde, (de) secours
based on	basé / fondé sur
blank space	espace
boot error	erreur d'initialisation
calculating machine	machine à calculer
calculator	calculette
calm *(v.)*	calmer
capacitor	condensateur
capital letter	majuscule
caps lock	majuscule verrouillée
check *(v.)*	vérifier
complete *(v.)*	compléter, achever
crash	incident
currently	actuellement
decade	décennie
default settings	réglage / paramétrage par défaut
delete *(v.)*	effacer
depend on *(v.)*	dépendre de
development	développement, évolution
dot	point
either ... or	ou … ou
enable *(v.)*	permettre, valider
energy consumption	consommation d'énergie
enter *(v.)*	entrer
equipped with	équipé de
escape	s'échapper
essential	essentiel
experience *(v.)*	expérimenter, vivre
feature	caractéristique
file	fichier
flicker-free	sans scintillement
floppy disk	disquette
font	police de caractères
frame	cadre
general	général
guarantee *(v.)*	garantie
hardware	machines
help desk	hotline, assistance en ligne
hesitate *(v.)*	hésiter
impetus	élan, force d'impulsion
ink jet (printer)	(imprimante) à jet d'encre
Insert	Insérer
integrated circuit	circuit intégré
interconnection	interconnection
invent *(v.)*	inventer
inventor	inventeur
key	clé, touche
large-scale (production)	à grande échelle
leaf through *(v.)*	feuilleter
lesson	leçon
mainly	principalement
meet someone's needs	répondre aux besoins de quelqu'un
monitor	moniteur
mousepad	tapis de souris
movie	film
(natural) science	science (naturelle)
notepad	cahier
overwrite *(v.)*	surfrapper
paper jam	bourrage papier
patch	modification de programme
performance	performance
pick up *(v.)*	ramasser
plug	fiche, prise
propose *(v.)*	proposer
purpose	but, objectif
recognise *(v.)*	reconnaître
reduce *(v.)*	réduire
reinstall *(v.)*	réinstaller
representative	représentant
required	demandé, exigé, requis
resistor	résistance
retailer	détaillant
retrieve data	récupérer des données
revolutionise *(v.)*	révolutionner
save *(v.)*	sauver, sauvegarder
scientist	scientifique
screech *(v.)*	hurler, grincer
shop assistant	vendeur

silicon wafer	tranche de silicium
Space bar	Barre espace
spin *(v.)*	(faire) tourner / pivoter
stack	tas, pile
storage capacity	capacité de stockage
store *(v.)*	stocker
suitable	approprié, qui convient
switch on *(v.)*	allumer
system disk	disque système
the file is corrupted	le fichier est altéré
troubleshooting	dépannage
unit	unité
update *(v.)*	mettre à jour
vacuum tube	tube électronique / à vide
weigh *(v.)*	peser
witness *(v.)*	témoigner de
word processor	machine de traitement de texte
write protection	protection en écriture

B

advantage	avantage
after-sales service	service après-vente
angle	angle
application software	logiciel d'application
appointment	rendez-vous
assembly line	chaîne de montage
assure *(v.)*	assurer
attachment	pièce jointe, fichier joint
bar graph	histogramme
billing	facturation
brochures	brochures, documentation
calculate *(v.)*	calculer
character	caractère
clipboard	calepin
column	colonne
combine *(v.)*	combiner
command	commande
correct *(v.)*	corriger
corresponding	correspondant
create *(v.)*	créer
customer service	service clientèle
cut-and-paste	couper-coller
database	base de données
deinstallation	désinstallation
design *(n., v.)*	conception ; créer, concevoir
desktop organiser / organizer	agenda électronique de bureau
dialogue box	boîte de dialogue
dictionary	dictionnaire
display *(v.)*	afficher
drag and drop	glisser et déposer
edit *(v.)*	éditer
efficiency	efficacité
efficient	efficace
enable *(v.)*	permettre
entire	entier
entry	entrée
estimate *(v.)*	estimer, faire une estimation
font size	taille de police
footer	titre de bas de page
for instance	par exemple
format *(v.)*	formater
gap	écart
grammar checker	vérificateur de grammaire
graph	graphique
graphic tool	outil graphique
header	en-tête
height	hauteur
hyphenation	coupure de mot
icon	icône
image	image
increase *(v.)*	augmenter
insert *(v.)*	insérer
instantly	instantanément
integrate *(v.)*	intégrer
interest rate	taux d'intérêt
internal operation	opération interne
invoice	facture
keep track of *(v.)*	suivre
line graph	courbe
load *(v.)*	charger
locate *(v.)*	localiser
machining	fabrication à la machine
manipulate *(v.)*	manipuler
manual	manuel
manually	manuellement
manufacturing unit	unité de fabrication
materials in stock	matériaux en stock
message	message
misspelling	faute d'orthographe, coquille
option	option
percentage	pourcentage
perform *(v.)*	effectuer, opérer, accomplir
pie chart	camembert
proofreading	correction d'épreuves
receive *(v.)*	recevoir
rectangle	rectangle
refined	raffiné
replace *(v.)*	remplacer
report	rapport
restart *(v.)*	recommencer, redémarrer

retype *(v.)*	retaper, resaisir
rotate *(v.)*	tourner, (faire) pivoter
row	rangée
save as *(v.)*	enregistrer sous
scale	échelle
search *(v.)*	rechercher
select *(v.)*	sélectionner
shape	forme
shipment	cargaison, envoi, expédition
skilled	qualifié
spell checker	vérificateur d'orthographe
spreadsheet program	tâbleur
square	carré
suggestion	suggestion, proposition
table	table, schéma
task	tâche
technical drawing	dessin technique
tedious	ennuyeux, fastidieux
thus	ainsi, donc
turnover	chiffre d'affaires
undo *(v.)*	défaire
uninstall *(v.)*	désinstaller
value	valeur
via	via, par
vice versa	vice versa
view	vue
width	largeur

C

above	au-dessus (de)
accessibility	accessibilité
accompany *(v.)*	accompagner
administrative offices	bureaux administratifs
alloy	alliage
animated	animé
aspect	aspect
astounding	étonnant
audience	public
availability	disponibilité
average	moyen(ne)
avoid *(v.)*	éviter
bank account	compte bancaire
beamer	protecteur
boldface	gras
cell(ular) phone *(US)*	téléphone portable
client	client
common	fréquent, commun
conquer *(v.)*	conquérir, vaincre
consumer report	sondage des consommateurs
defence system	système de défense
destination	destination
destroy *(v.)*	détruire
disadvantage	inconvénient
domain	domaine
download *(v.)*	télécharger
drown *(v.)*	noyer
enterprise	entreprise
exaggerate *(v.)*	exagérer
finding	résultat, conclusion
foamed	écumé
gather *(v.)*	cueiller, recueillir
global	global, mondial
guide *(v.)*	guider
handle *(v.)*	traiter
hardly	ne … guère
hero	héros
heroine	héroïne
hint	allusion, indice
hypertext	hypertexte
immense	immense
impression	impression
in addition	en plus
include *(v.)*	inclure
incorporate *(v.)*	incorporer
information management	gestion d'information
interrupt *(v.)*	interrompre
involve *(v.)*	comprendre, consister en
item	article
keyword	mot clé
lawyer	avocat
legal	juridique, légal
melting point	point de fusion
mention *(v.)*	mentionner, citer
mobile phone *(UK)*	téléphone portable
noise	bruit
nowadays	de nos jours
outdated	démodé
path	chemin
precautions	précautions
prefer *(v.)*	préférer
pronunciation	prononciation
protect *(v.)*	protéger
refer to *(v.)*	faire référence à, se référer à
regarding	concernant
reliability	fiabilité
reliable	fiable
rely on *(v.)*	compter sur, faire confiance à
reprint *(v.)*	réimprimer
research	recherche(s)
safety	sécurité
search engine	moteur de recherche
serious	sérieux, grave
server	serveur
shareholder	actionnaire
soccer club	club de football
source	source

special-interest magazine revue spécialisée
spy *(v.)* espionner
surely sûrement
the same holds if … il en est de même si …
thermosetting plastics plastiques thermodurcissables
unlicensed sans permis, non autorisé
up-to-date moderne
up-to-dateness modernité
virus protector dispositif de protection contre les virus
volume volume

UNIT 4 • Starter

adjustable wrench clé à molette
Allen key(s) / hexagon key clé mâle
bend *(v.)* plier
bolt boulon
bore *(v.)* percer
centre lathe tour à pointes
clamp *(v.)* serrer, cramponner
column drill perceuse à colonne, fraiseuse
combination pliers pince universelle
debur *(v.)* ébarber, ébavurer
diameter diamètre
drill *(v.)* percer
drive in *(v.)* enfoncer
(electric) drill perceuse (électrique)
equipment matériel
file *(v.)* limer
hacksaw scie à métaux
indispensable indispensable
join *(v.)* joindre, attacher
loosen *(v.)* déserrer
machine tool(s) machine-outil(s)
mark *(v.)* marquer
measure *(v.)* mesurer
metal bar barre de métal
milling machine fraiseuse
nail *(n., v.)* clou ; clouer
nut écrou
operate *(v.)* opérer
power tool(s) outil électrique
riveting tongs pince à rivets
screw vis
screwdriver tournevis
shaft arbre, manche, tige
spanner clé
split *(v.)* fendre
stripping tongs pince à dénuder
tighten *(v.)* serrer
toolbox boîte à outils
tools outils
vice étau
work *(v.)* travailler

A

adhesive adhésif
apply *(v.)* appliquer
apprentice apprenti
assemble *(v.)* assembler
basic training formation de base
bond *(v.)* lier, unir
capacity capacité
ceramic(s) céramique
chief executive directeur général
Computerized Numerical Control (CNC) commande numérique
competence compétence
demonstrate *(v.)* démontrer, présenter
depend on *(v.)* dépendre de
design *(n., v.)* plan, conception, dessin ; concevoir, dessiner
determine *(v.)* déterminer
Do-It-Yourself (DIY) bricolage
emergency urgence
engineering génie
exhaust pipe pot d'échappement
experience expérience
general workshop atelier
(hexagon) socket screw vis à tête hexagonale
in spite of malgré
include *(v.)* inclure
fair foire, salon
join *(v.)* joindre, attacher
manufacture *(n., v.)* fabrication ; fabriquer
mechanical engineering mécanique
operate a machine *(v.)* faire fonctionner une machine
perform *(v.)* effectuer, opérer, accomplir
performance performance
plastics plastiques
prompt rapide, prompt
property propriété
rivet *(n., v.)* rivet ; riveter
rpm (revolutions per minute) tr/min (tours à la minute)
screw vis
sheet steel tôle d'acier
shop floor atelier
sign panneau
skilled worker ouvrier qualifié
solder *(v.)* souder
soldering soudure
spindle arbre, broche, axe

state-of-the-art	de pointe
steel plate	plaque d'acier
tool kit	jeu d'outils
train *(v.)*	former
weld *(v.)*	souder
welding	soudure
welding torch	chalumeau
workbench	banc de travail, établi

B

abrasive	abrasif
accessory	accessoire
accuracy	précision
adjustment	réglage
advantage	avantage
advertise *(v.)*	faire de la publicité, passer une annonce
advertisement	annonce, (une) publicité
alter *(v.)*	modifier, changer
angle grinder	disqueuse
aspect	aspect
auto shut-off	arrêt automatique
band saw	scie à ruban
base	base
base table	table
bearings	palier
bed	base, bâti
belt	courroie
bit	foret
cartridge	cartouche
chainsaw	tronçonneuse
charge *(v.)*	charger
charger	chargeur
chest of drawers	commode
chuck	mandrin
clamping system	système de blocage
clip	collier, bague, pince
clockwise	dans le sens des aiguilles d'une montre
clutch	embrayage
column	colonne
compartment	compartiment
control *(n., v.)*	contrôle ; contrôler, régler
control panel	panneau de contrôle
convenient	commode
cordless	sans fil
counter- / anti-clockwise	en sens inverse des aiguilles d'une montre
cross-slide	chariot longitudinal
DC (direct current)	courant continu
depth gauge	calibre de profondeur
device	appareil
discharged	déchargé
discount	remise
discount store	solderie
dispose of *(v.)*	se débarrasser de
domestic	domestique
ecological	écologique
emergency stop	arrêt d'urgence
essential	essentiel
extensive	important, considérable
extractor	extracteur, ventilateur
feature	caractéristique, particularité
feed *(v.)*	alimenter
frame	cadre
gear	engrenage, vitesse
gear and control unit	commande de variateur de vitesses
gearbox	boîte de vitesses
gears	vitesses
goggles	lunettes de travail
grinder	rectifieuse, meuleuse, pileur, broyeur
grinding machine	rectifieuse, affûteuse
grinding wheel	disque abrasif
handle *(n., v.)*	manche ; manier
have ... in common	avoir … en commun
headstock	poupée (d'un tour)
house *(v.)*	recevoir, contenir
impact drill	perceuse à percussion
improvement	amélioration
indicate *(v.)*	indiquer
instruction	instruction
insulation	isolement
jigsaw	scie anglaise
keyless chuck	mandrin sans clé
knee	capot
large-scale manufacturing	fabrication à grande échelle
latest	dernier
leaflet	dépliant
lever	levier
light-duty	léger
lock out	blocage
locking lever	levier de blocage
longitudinal	longitudinal
low	bas
main switch	interrupteur principal (marche-arrêt)
major benefit	bénéfice principal
motor-reversing switch	interrupteur pour inverser le moteur
mount *(v.)*	monter, installer
mud guard	garde-boue
overarm	traînard
plug	fiche, prise
polish *(v.)*	polir
preselection	présélection

present *(v.)*	présenter
provide *(v.)*	fournir
range	gamme
recharge *(v.)*	recharger
rechargeable	rechargeable
recyclable	recyclable
reliable	fiable
require *(v.)*	demander, exiger, requérir
required	demandé, exigé, requis
reset *(v.)*	réenclencher, réinitialiser
reversible	reversible
revolution speed	vitesse de révolution
rotate *(v.)*	tourner, (faire) pivoter
saddle	selle
safety	sécurité
sander	ponceuse
sense of rotation	sens de rotation
slideway	coulisse
socket	godet, cosse, emboîtement
speed-selection lever	levier pour sélectionner la vitesse
starting switch	commande marche-arrêt
storage	stockage
strength	force
superior	supérieur
support *(v.)*	supporter, soutenir, maintenir
table	table
table hand feed	table à alimentation manuelle
table traverse	table transversale
tailstock	contre-poupée
trigger switch	gâchette
toolpost	chariot porte-outils
torque	moment de torsion
transmit *(v.)*	transmettre
variable	variable
various	divers
vertical milling machine	fraiseuse verticale
vertical slide	coulisse verticale
voltage (V)	tension

C

appropriate	approprié, qui convient
aware of	conscient de
border	bordure
crossbar	barre de traverse d'accouplement
emergency	urgence
emergency sign	panneau
escape route	voie de secours, sortie de secours
exit	sortie
fire extinguisher	extincteur
fire sign	panneau : risque d'incendie
first aid kit	trousse à pharmacie
forbid *(v.)*	interdire
gangway	passage
get injured	être blessé
glove	gant
guard	dispositif de protection
handshield	protège-main
label	étiquette
mains	secteur
mandatory sign	panneau obligatoire
prohibition	interdiction
protective	de protection
route	itinéraire, chemin
safety boots	chaussures de sécurité
safety sign	panneau de sécurité
triangular	triangulaire
vital	essentiel

UNIT 5 • Starter

annoying	embêtant
auxiliary handle	manche auxiliaire
blocked	bloqué
cracked	fendu
jammed	bloqué
blunt	peu tranchant
figure out *(v.)*	arriver à comprendre
hole	trou
housing	capot
keyless chuck	mandrin sans clé
maintenance	maintenance
required	demandé, exigé, requis
sharpen *(v.)*	aiguiser
troubleshooting	dépannage
twist drill	perceuse à spirale
warranty	garantié

A

about to	sur le point de
diagram	diagramme
give sombody a ring	passer un coup de fil à quelqu'un
high speed steel (HSS)	HSS
locked in	(ici) engagé
master	maître
ratio	taux
release *(v.)*	relâcher

B

application	application
available	disponible
basic rule	règle de base
brad point	pointe à tête perdu
common	commun
concerned	concerné
concrete	béton
cope with *(v.)*	s'occuper de
craftsman	artisan
determine *(v.)*	déterminer
diameter (dia)	diamètre
Do It Yourself (DIY)	bricolage
drill bit	foret de perceuse
due to	à cause de
e.g. (= for example)	par exemple
expand *(v.)*	agrandir, développer
however	pourtant
increase *(v.)*	augmenter
instruction	instruction
manual	manuel
mention *(v.)*	mentionner, faire mention de, citer
multi-purpose	à usages multiples, polyvalent
packaging	conditionnement
popular	populaire
properties	propriétés
recent	récent
recommend *(v.)*	recommander
relevant	pertinent
select *(v.)*	sélectionner
state *(v.)*	déclarer, dire
tungsten-carbide	carbure de tungstène
versatile	polyvalent

C

accident	accident
adhesive	adhésif
affect *(v.)*	avoir un effet sur
appropriate	approprié, qui convient
apron	tablier
asbestos	amiante
attached	attaché, joint
authorized	autorisé
brass	laiton
brick	brique
care	soin
carry out *(v.)*	effectuer
cast iron	fonte
ceiling	plafond
cement	ciment
centre-punch	pointeau, amorçoir
claim	réclamation
clamping jaw	mordache
component	composant
countersunk screws	vis noyées
crystalline	cristallin
current surge preventing switch	empêcher les à-coups de courant interrupteur
damage *(v.)*	endommager
defect	défaut
distort *(v.)*	distordre
double insulation	isolement double
drill the surface	percer la surface
exploded view	vue éclatée
extend *(v.)*	(s')étendre, (se) développer
failure	échec
faulty	défectueux
frequent	fréquent
harmful to the health	nuisible à la santé
impact drilling	perçage à percussion
improper	inadéquat, mauvais
instead	à la place
keep clear *(v.)*	(se) mettre à l'écart
lead	extension, rallonge
lubricant	lubrifiant
lubricate *(v.)*	lubrifier
lubrication	lubrification
marble	marbre
masonry	maçonnerie
minor	mineur
mishandling	mauvais usage
misuse	mauvais usage, usage abusif
modification	modification
occur *(v.)*	se passer
overheating	échauffement
paraffin	pétrole, mazout
percussion drilling	perçage à percussion
performance	fonctionnement
pierce *(v.)*	percer
pilot hole	avant-trou
pre-drill *(v.)*	pré-percer
prepay *(v.)*	prépayer
prevent *(v.)*	empêcher
proof	preuve(s)
purchase	achat
range of accessories	gamme d'accessoires
receipt	reçu
recommendation	recommandation
render *(v.)*	rendre
replace *(v.)*	remplacer
replacement	remplacement
reversing lever	levier d'inversion de marche
sawdust	sciure
service	service

sheet metal	métal en feuilles, tôle
shipment	cargaison, envoi, expédition
silicic acid	acide silicique
slot	fente
solder *(v.)*	souder
spare part	pièce de rechange, pièce détachée
spiral drill bit	foret en spirale
splinter	éclat
statement	déposition, déclaration
strain	charge
sturdy	solide
supply *(v.)*	fournir
surge	à-coup
tile	tuile
tip	bout, pointe
turpentine	térébenthine
ventilation	ventilation
warranty	garantie
wear	usage
working range	zone de travail

UNIT 6 • Starter

appliance	appareil
bulb	ampoule
device	appareil
manual	manuel
motion	mouvement
portable	portable, portatif
wall socket	prise murale

A

alternating current	courant alternatif
alternator	alternateur
ammonium chloride	chlorure d'ammonium
amp	ampère
carbon	carbone
coil	bobine
combination pliers	pince universelle
commutator	commutateur
cross slot screw	vis à tête cruciforme (Phillips)
current	courant
detergent	détergent
direct current	courant continu
discharge *(v.)*	décharger
dissolve *(v.)*	dissoudre
door knob	poignée de porte
electric charge	charge électrique
electrostatic induction	induction électrostatique
employ *(v.)*	employer, utiliser
fatal	mortel
faulty	défectueux
frequency	fréquence
friction	friction
fuse	fusible
generate *(v.)*	générer
hertz (Hz)	hertz
immerse *(v.)*	immerger
injured	blessé
jumper	pullover
lightning	éclair
live conductor	conducteur sous tension
live wire	fil sous tension
mains testing screwdriver	tournevis testeur
mouth-to-mouth	bouche-à-bouche
plug	fiche, prise
respiration	respiration
regardless	quand même
remedy	remède
repel *(v.)*	repousser
respiratory failure	défaillance respiratoire
rod	barre
screwdriver	tournevis
slot screw	vis à tête fendue
soldering iron	fer à souder
spin *(v.)*	(faire) tourner / pivoter
tap	taper, tapoter
to and fro	mouvement de va-et-vient
torch	lampe
vice-versa	vice versa
voltage tester	testeur de tension
wire stripper	pince à dénuder
zinc	zinc

B

brush	brosse
connect *(v.)*	connecter
connected in parallel	connecté en parallèle
connected in series	connecté en série
discharging process	processus de déchargement
lead	plomb
link together *(v.)*	lier, joindre
recharging process	processus de rechargement
repulsive force	force répulsive
reverse *(v.)*	inverser
rotor	rotor
stack	tas, pile
stator	stator
sulphuric acid	acide sulfurique

C

abundant abondant
electrolyte électrolyte
efficiency efficacité
emission émission
fuel cell pile à combustible
hydrogen hydrogène
in the long term à long terme
internal combustion engine (ICE) moteur à combustion interne
pollutant polluant
preferable préférable
propulsion propulsion
range gamme
thus ainsi, donc
water vapour vapeur d'eau
well puits
zero emission vehicle voiture à émission(s) zéro

D Grammar revision

catalytic converter pot catalytique
harmful nuisible, nocif
lamp socket douille
sophisticated complexe, perfectionné

UNIT 7 • Starter

aluminium *(UK)*, **aluminum** *(US)* aluminium
ceramics céramique
copper cuivre
fibreglass fibre de verre
iron fer
lead plomb
plastics plastiques
property propriété
steel acier
wood bois

A

adjust *(v.)* régler
advantage avantage
appliance appareil
bookend serre-livres
choice choix
circle cercle
circular circulaire, rond
consist of *(v.)* consister en
cord cordon, câble
cube cube
desk tidy organiseur de bureau
diameter diamètre
disadvantage inconvénient
ellipse ellipse
figure figure
focus *(v.)* focaliser, mettre au point, fixer, centrer, concentrer
furniture mobilier
general général
glue *(v.)* coller
height hauteur, taille
housing capot
keyword mot-clé
label *(v.)* étiqueter
length longueur
lens lentille, objectif, verre
measurement mesure
mention *(v.)* mentionner, faire mention de, citer
mirror miroir, glace
old-fashioned démodé
oval ovale
polygon polygone
prepare *(v.)* préparer
presentation présentation
project *(v.)* projeter
radius rayon
reason raison
rectangle rectangle
rectangular rectangulaire
reflector réflecteur
right angle angle droit
rigid rigide
screen écran
screw *(v.)* visser
shape forme
shiny brillant
size taille
specific précis
square carré
state-of the-art de pointe
sturdy solide
surface surface
switch interrupteur
timeless intemporel, hors du temps
transformer transformateur
transparency transparent
triangle triangle
weight poids
weld *(v.)* souder
wheel roue
width largeur
cylinder cylindre

B

acid	acide
addition	addition
alloy	alliage
amount	montant, nombre, quantité, somme
and the like	et d'autres choses comme cela
apart from	à part
art	art
artificial	artificiel
ball bearing	bille de roulement
base	base
BC (= before Christ)	avant J.-C.
brass	laiton
brittle	cassant, fragile
brittleness	fragilité
bronze	bronze
bucket	seau
casing	revêtement
cast *(v.)*	couler, fondre
cast iron	fonte
chart	schéma, diagramme
chromium	chrome
clay	argile
component	composant
composite	composite
compound	composé
conclusion	conclusion
concrete	béton
conductivity	conductivité
confuse *(v.)*	confondre
consequently	par conséquent
corrode *(v.)*	corroder
corrosion	corrosion
crown	couronne
cutting tool	outil de découpage / cisaillage
dental	dentaire
dentist	dentiste
depend on *(v.)*	dépendre de
disposal	enlèvement, traitement
dispose of *(v.)*	se débarrasser de
divide *(v.)*	diviser
durability	durabilité
durable	durable, résistant
elasticity	élasticité
employ *(v.)*	employer, utiliser
enduring	durable
engineering	génie
fairly	assez
ferrous metals	métaux ferreux
fumes	exhalaisons, fumées
furnace	fourneau
goldsmith	orfèvre
granulate	granulé
handout	prospectus, polycopié
harden *(v.)*	durcir
hardness	dureté
improve *(v.)*	(s')améliorer
indefinitely	indéfiniment
indispensable	indispensable
ingredient	ingrédient
inlay	incrustation
inorganic	inorganique
instead of	au lieu de
intend *(v.)*	avoir l'intention de
jewellery	bijoux
joint	joint
knowledge	connaissances
leather	cuir
lime	chaux
lower *(v.)*	baisser
manganese	manganèse
melt *(v.)*	(faire) fondre
melting point	point de fusion
mild steel	acier doux
mould *(v.)*	mouler
nitrogen	azote
numerous	nombreux
ore	minerai
overview	vue d'ensemble
oxide	oxyde
paper clip	trombone
permanent	permanent
phosphorus	phosphore
plentiful	abondant
poisonous	vénéneux, toxique
polyamide	polyamide
porcelain	porcelaine
pottery	poterie, céramiques
pressure	pression
processing	traitement
pure	pur
purpose	but, objectif
raw material	matière première
refer to *(v.)*	faire référence à
reinforced	renforcé
re-mould *(v.)*	remouler, rechaper
request *(v.)*	demander
require *(v.)*	demander, exiger, requérir
reshape *(v.)*	reformer
resistance	résistance
resistant to	résistant à
reuse *(v.)*	réutiliser
rust *(v.)*	rouille
sheet steel	tôle d'acier
silicon	silicium
sink	évier
sintered metal	métal fritté
soften *(v.)*	ramollir, adoucir, assouplir

softness	douceur, moelleux
source	source
spelling	orthographe
stainless steel	acier inoxydable, Inox
suitable	approprié, qui convient
sulphur	soufre
support *(v.)*	supporter, soutenir, maintenir
susceptible to	prédisposé à
synthetics	synthétiques
teeth *(pl.)*	dents
term	condition, terme
thermoplastic	thermoplastique
thermosetting (plastics)	(plastiques) thermodurcissables
thorough	minutieux, approfondi
toy	jouet
tungsten	tungstène
utility	utilité
valve	vanne
waste	déchets
weak	faible

C

abrasion	abrasion
accurate	précis
achieve *(v.)*	réaliser
affect *(v.)*	avoir un effet sur
aligned	aligné
apologise *(v.)*	s'excuser, regretter
apply for *(v.)*	poser sa candidature pour, faire une demande de
approximately	approximativement
audit *(v.)*	vérifier, apurer
auxiliary scale	échelle auxiliaire, micromètre
available	disponible
axle	axe, essieu
base unit	unité de base
bottom	fond
calibrate *(v.)*	calibrer
capable	capable
carpenter	menuisier
clamp screw	vis d'arrêt, vis de blocage
conductive	conducteur
continual	continu
craftsman	artisan
current	courant
cutting edge	coupant, tranchant
decision	décision
deform *(v.)*	déformer
depth gauge	calibre de profondeur
develop *(v.)*	(se) développer
dimension	dimension
door lock	serrure de porte
drop *(v.)*	laisser tomber
ductility	ductilité, malléabilité
elasticity	élasticité
emphasis	accent
establish *(v.)*	établir
external measurement	mesure / dimension extérieure
face *(v.)*	être en face de
fasten *(v.)*	attacher
feeler gauge	jauge d'épaisseur / de profondeur / de palpage
film speed	vitesse de film
fixed	fixé
fixed jaw	mâchoire fixe
folding rule	mètre pliant
freight	fret
guideline(s)	ligne directrice, directives
hollow	creux
implement *(v.)*	implémenter
inside measurement	mesure / dimension intérieure
insist on *(v.)*	insister sur
inspection	inspection
internal measurement	mesure / dimension intérieure
introduction	introduction
isolated	isolé
issue *(v.)*	émettre
jaw	mâchoire
joiner	menuisier
large	grand
large-scale	à grande échelle
lathe	tour
layout	maquette, disposition
machining	usinage
main scale	échelle principale (en millimètres)
major	majeur
malleability	malléabilité
manufacture	fabrication
meet the requirements	satisfaire aux exigences
merge *(v.)*	fusionner
micrometer	micromètre
mov(e)able	mobile
nonius	micromètre
outside measurement	mesure / dimension externe
plant	usine
platinum	platine
plumber	plombier
precision	précision
prevent *(v.)*	empêcher

primarily	principalement
procedure	procédure
production output	production
prove *(v.)*	prouver
quality assurance	garantie de qualité
recognise *(v.)*	reconnaître
regarding	concernant
registration	inscription
representative	représentant
resources	ressources
result	résultat
review *(v.)*	examiner, revoir
scale	échelle
sliding caliper	calibre à coulisse
spark plug	bougie
spring	ressort
staff	personnel
steel square	équerre en acier
substance	substance
summary	sommaire
supplier	fournisseur
tape measure	mètre
thickness	épaisseur
tube	tube
unit	unité
utmost	le plus grand
vernier caliper	jauge micrométrique
versatile	polyvalent
workpiece	pièce

UNIT 8 • Starter

according to	selon
bar chart	histogramme
hard	dur
hardly	à peine, ne … guère
line graph	courbe
renewable energy	énergie renouvelable
sharp rise	forte augmentation, montée en flèche
slight	léger

A

a large extent	grande mesure
adjust *(v.)*	régler
aerosol	aérosol
alter *(v.)*	modifier, changer
anomaly	anomalie
attribute to *(v.)*	attribuer à
average temperature	température moyenne
carbon dioxide	dioxyde de carbone
cattle	bétail
CFC gases	gaz CFC
chlorofluorocarbon	chlorofluorocarbone
coolant	caloporteur, liquide de refroidissement
decade	décennie
derive *(v.)*	provenir, venir
disastrous	désastreux
disease	maladie
fertilizer	engrais
glacier	glacier
mean temperature	température moyenne
methane	méthane
nitrous oxide	oxyde nitreux
pattern	modèle, motif, système
pollution	pollution
prohibit *(v.)*	interdire
propellant	propulsant, carburant
sea level	niveau de la mer

B

achieve *(v.)*	réaliser
approximately	approximativement
battery charger	chargeur de pile / batterie
curved	courbé
environmentally-friendly	respectueux de l'environnement
estuary	estuaire
float	flotte
gadget	gadget
harness *(v.)*	harnais
high tide	marée haute
hydroelectric power	énergie hydroélectrique
inverter	inverteur
kinetic	kinétique
kinetic energy	énergie kinétique
liquid	liquide
low tide	marée basse
peak hour	heure de pointe
photovoltaics	photovoltaïque
pocket calculator	calculette
regard *(v.)*	considérer
reliable	fiable
semiconductor	semiconducteur
site	site
solar furnace	four solaire
solar panel	panneau solaire
technology	technologie
tidal power	énergie marémotrice
tide	marée
unpredictable	imprévisible
warranty	garantie
waste	déchets
wave power	énergie des vagues

C

abundance	abondance
cancer	cancer
detrimental	nuisible, préjudiciable
deuterium	deutérium
fossil fuels	combustibles fossiles
fuse together *(v.)*	fusionner
excess	excès, surplus
more likely	plus probable
nuclear fission	fission nucléaire
nuclear fusion	fusion nucléaire
nuclear waste	déchets nucléaires
nucleus	noyau
pellet	boulette
vessel	vaisseau

UNIT 9 • Starter

travel arrangements	organisation de / préparatifs pour le voyage

A

accept *(v.)*	accepter
access	accès
advise *(v.)*	conseiller
advisor	conseil, conseiller
aisle	allée
apologise *(v.)*	s'excuser
arrange *(v.)*	organiser
arrival	arrivée
attraction	attraction
available	disponible
avoid *(v.)*	éviter
booked up	complet
booth	cabine
branch	branche, agence
car rental	location de voiture
chamber of commerce	chambre de commerce
check *(v.)*	vérifier
check-in	enregistrement, arrivée, accueil
check-out	heure de départ (hôtel)
circumstance	circonstance
Columbus Day *(US)*	jour commémorant l'arrivée de Christophe Colomb en Amérique (2e lundi d'octobre)
commemorative	commémoratif
complete *(v.)*	compléter, achever
complimentary	gratuit, gracieux
concerning	concernant
confirm *(v.)*	confirmer
connecting rooms	chambres communicantes
convenient	bien situé
courtesy	gratuit, service offert aux clients
create *(v.)*	créer
crib	lit d'enfant
customer	client
damp	humide
decide on *(v.)*	se décider pour
delivery	livraison
depart *(v.)*	partir
departure	départ
domestic flight	vol intérieur
double room	chambre double
downtown *(US)*	centre ville
driver's licence	permis de conduire
dry cleaning	pressing
economy class	classe touriste
employer	employeur
environment	environnement
equipment	équipement, matériel
establish *(v.)*	établir
exercise gym	salle de gym
expect *(v.)*	s'attendre à
express	express
facilities	installations, équipements
fair *(n.)*	foire, salon
features	caractéristiques, particularités
finalize *(v.)*	finaliser
foam pillows	oreillers en mousse
(be) fond of	aimer
function space	salle de réception
gain a foothold *(v.)*	prendre pied
genius	génie
grant *(v.)*	accorder
guess *(v.)*	deviner
hairdryer	sèche-cheveux
hand luggage	bagage à main
hold the line	ne pas quitter
in addition	en plus
include *(v.)*	inclure
Independence Day *(US)*	fête nationale de l'Indépendance (le 4 juillet)
indicate *(v.)*	indiquer
iron	fer à repasser
ironing board	planche à repasser
jacuzzi	jacuzzi
Labo(u)r Day *(US)*	fête du travail (1er lundi de septembre)
laundry	buanderie
local	local

located	situé
major	majeur, important
membrane keyboard	clavier avec bâche de protection
membrane switch	interrupteur avec bâche de protection
Memorial Day *(US)*	jour férié en l'honneur des soldats américains morts pour la patrie (dernier lundi de mai)
misunderstanding	malentendu
newsstand	kiosque à journaux
non-allergenic	hypoallergénique
offer *(v.)*	offrir
personnel	personnel
pick someone up *(v.)*	aller chercher quelqu'un (en voiture)
plant	plante
pool	piscine
postal service	service des postes
prefer *(v.)*	préférer
present *(v.)*	présenter
pressing	pressing
promote *(v.)*	promouvoir
prompt	rapide, prompt
property	propriété
rate	taux, prix
receptionist	réceptionniste
refrigerator	réfrigérateur
represent *(v.)*	représenter
request *(v.)*	demander
row	rangée
schedule	planning, horaire
security check	contrôle de sécurité
settle *(v.)*	résoudre (un problème)
shoe shine	cirage des chaussures
shower	douche
shuttle	navette
single room	chambre simple
stocked	stocké
stopover	escale
suburban	de banlieue
successful	réussi
suggestion	suggestion, proposition
suit *(v.)*	convenir à
suitable	approprié, qui convient
sundries	articles divers, extras
Thanksgiving Day *(US)*	fête nationale qui commémore l'installation des premiers colons (4e jeudi de novembre)
ticket agent	agence de spectacles / de voyages
travel agent/agency	agence de voyages
vacation *(US)*	vacances
valet service	pressing (de l'hôtel)
Veterans Day *(US)*	le 11 novembre

B

admission	entrée
air conditioning	climatisation
anniversary	anniversaire
announcement	annonce
board	tableau
cast *(v.)*	couler, fondre
celebrate *(v.)*	fêter
charter	charte
colonial age	époque coloniale
compact (type of car)	petite voiture
crack	fissure
debate	débat
design *(v.)*	concevoir, dessiner
exhibit *(v.)*	exposer
explanation	explication
family name	nom de famille
favourable	favorable
female	féminin
first name	prénom
flight attendant	steward, hôtesse de l'air
full-size (type of car)	grosse voiture
gather *(v.)*	cueillir, recueillir
historic square mile	centre historique
imagine *(v.)*	imaginer
immigration	immigration
immigration card	carte d'immigration
information sheet	feuille d'information
inn	auberge
inscribe *(v.)*	inscrire
male	masculin
marital status	état civil
middle name	deuxième prénom
mid-size (type of car)	voiture de taille moyenne
narrow	étroit
passenger	passager
peaceful	paisible, tranquille
privilege	privilège
provide *(v.)*	fournir
recast *(v.)*	refondre
recommend *(v.)*	recommander
relax *(v.)*	se détendre
return time	heure de retour
revolutionary	révolutionnaire
screen	écran
sew *(v.)*	coudre
sign *(v.)*	signer
straight	droit
surname	nom de famille
trade guild	organisme professionnel
traveller	voyageur
waiter	serveur
waitress	serveuse
wherever	n'importe où

C

amount to *(v.)* s'élever à
asparagus asperge
beef bœuf
beverage boisson
bill *(UK)* addition
brook petite rivière
check *(US)* addition
chef chef (de cuisine)
chilli chili
chutney chutney (condiment à base de fruits)
complemented complété, accompagné
cordial cordial
cornmeal farine de maïs
crab cake croquette de crabe
cranberry airelle
cream crème
crisp croquant
cucumber concombre
dessert dessert
domestic national
draft / draught beer bière pression
duckling caneton
entrée *(US)* plat principal
first course entrée
fried frit, poêlé
game gibier
giant géant
glazed glacé
herbed avec des herbes
herbs herbes
horseradish raifort
jumbo géant
lamb agneau
lemon citron
lemongrass citronnelle
lobster homard
main course plat principal
medium à point (steak)
menu carte
mushroom champignon
mustard moutarde
oyster huître
pancake crèpe
peach pêche
pecan pecan
pepper poivre
pewter étain
pickapeppa remoulade *(US)* mayonnaise à la paprika
poultry volaille
prime rib côte de bœuf
puff pastry pâte feuilletée
rack of lamb carré d'agneau
rare saignant (steak)
roasted rôti, cuit au four
sauce sauce
seasonal de la saison, saisonnier
serve *(v.)* servir
service service
shallot échalote
sign panneau
smoked fumé
soup soupe, potage
spice *(v.)* épicer
spicy épicé
spinach épinards
spirits spiritueux
squash *(US)* courge
stuffed farci
subdivide *(v.)* subdiviser
take orders prendre des commandes
tip pourboire
top *(v.)* napper (sauce)
trout truite
veal veau
vegetable légume
vegetarian végétarien
venison venaison
walnut noix
well-done bien cuit (steak)

D Grammar revision

advice conseil(s)
bonus bonus, prime
success succès, réussite
traffic jam bouchon

UNIT 10 • Starter

distribution diffusion, distribution
enable *(v.)* permettre
goods biens, marchandises
grant *(v.)* accorder
national national
tremendously énormément

A

access (to) accès (à)
as far as … is concerned en ce qui concerne …
cost-intensive country pays où les coûts de production sont élevés
expectation attente
freelancer travailleur indépendant

highly skilled	très qualifié
intensify *(v.)*	intensifier
limitless	sans limites
low-cost country	pays où les coûts de production sont bas
means *(n. sing.)*	moyen

B

cohesion	cohésion
Department for Employment and Education *(UK)*	ministère britannique de l'Emploi et de l'Éducation
dependent child	enfant à charge
funds	fonds
Home Office *(UK)*	ministère britannique de l'Intérieur
obtain *(v.)*	obtenir
offer	offre
official	officiel, agent
penetration	pénétration
residence permit	permis de séjour
right of residence	droit de séjour
self-employed	indépendant, qui travaille à son compte
self-employed person	personne qui travaille à son compte
spouse	époux, épouse
sustained economic growth	croissance économique soutenue
trade union	syndicat
with respect to	en ce qui concerne
work permit	permis de travail

C

a plus	un plus
achievement	exploit, réussite
alert	alerte
animation skills	compétences d'animateur
basic CAD skills	compétences de base en CAO
benefit package	les avantages
capitalisation	capitalisation
car component supplier	fournisseur de composants pour l'industrie automobile
career prospect	perspective de carrière
confident	confiant
controversial issue	sujet de controverse
deadline	temps limite, délai
derogatory	dérogatoire
equivalent	équivalent
facility	facilité
focus *(v.)*	focaliser, mettre au point, fixer, centrer, concentrer
fringe benefit(s)	avantage(s) en nature
from conception to completion	de la conception à la réalisation
implementation	implémentation
in public	en public
inappropriate	inapproprié
minor thing	chose mineure
negotiable	à négocier
on file	sur fichier
optimal solution	solution optimale
panel discussion	débat, tribune
prove *(v.)*	prouver
reject *(v.)*	rejeter
request *(v.)*	demander
require *(v.)*	demander, exiger, requérir
sample	échantillon
schedule *(v.)*	planifier
suitable	approprié, qui convient
tough	dur

Reading Files • File 1
Civil engineering

access	accès
acoustic insulation	isolement acoustique
administration	administration
allowance	allocation
altering	modification
apartment block	immeuble d'habitation
(be) concerned with	concerner
beam	poutre, rayon
bear *(v.)*	supporter, soutenir
block	bloc
bond *(v.)*	(s')adhérer
brick	brique
brick course	assise
buckle *(v.)*	boucler
buttress	contrefort
ceiling	plafond
chimney stack	cheminée
civil engineer	ingénieur des travaux publics
civil engineering	génie civil
clay	argile
compact *(v.)*	compacter
compressive strength	force
concrete base	base / soutien en béton
concrete block	bloc de béton
connection	lien

corrosion	corrosion
damp	humide
dead load	poids mort
deep	profond
deflection	déflection
demand	demande
design	conception, dessin
distribution	distribution, diffusion
divide *(v.)*	diviser
drainage	système d'égouts / d'évacuation des eaux
emergency escape route	sortie de secours
enable *(v.)*	permettre
enclosure	pièce jointe
ensure *(v.)*	assurer
equivalent	équivalent
essential	essentiel
even	régulier, uniforme
excavate *(v.)*	excaver, creuser
exceed *(v.)*	dépasser
external	externe
finish	fin, terminaison
fir	sapin
fork lift truck	chariot à fourche, gerbeuse
foundation	fondation
frame	cadre
high-rise building	tour
hydraulics	hydraulique
impose *(v.)*	imposer
imposed load	charge imposée
intermediate	intermédiaire
joint	joint
joist	solive
landscape	paysage
larch	mélèze
lateral	latéral
live load	poids roulant
load	charge
load-bearing	porteur
load-bearing wall	mur porteur
loss	perte
masonry	maçonnerie
meet human requirements	satisfaire aux exigences humaines
obligatory	obligatoire
paramount	primordial
partition (wall)	cloison
penetration	pénétration
permanent	permanent
pier	pile
pine	pin
pitched roof	toit en pente
pre-cast concrete floor	sol en béton préfabriqué
prefer *(v.)*	préférer
pre-stressed concrete	béton préconstraint
prevent *(v.)*	empêcher
privacy	intimité
provision	disposition, provision réserve
purpose	but, objectif
push sideways *(v.)*	pousser de côté
reduce *(v.)*	réduire
redwood	séquoia
reinforced	renforcé
reinforced concrete	béton renforcé
requirement	exigence
residential	résidentiel
resist *(v.)*	résister à
restrict *(v.)*	restreindre, limiter
river control	contrôle fluvial
sanitation	système sanitaire, hygiène publique
sheet material	matériel en feuilles
shuttering	coffrage
soil	sol
span	portée, largeur
spread	étendue, développement
spread *(v.)*	s'étendre
spruce	épicéa
squash *(v.)*	écraser
staircase	escalier
storage	stockage
stretch *(v.)*	s'étendre
strip	bande, bandeau
structural calculation	calcul structural
structural form	forme structurale
subdivision	subdivision
subject to	assujetti à
suction	succion, aspiration
sufficient	suffisant
supply	fourniture, offre
support *(n., v.)*	support ; maintenir, soutenir, supporter
suspended floor	plancher suspendu
take into account *(v.)*	prendre en compte
tension	tension
tie *(v.)*	lier
timber	bois de construction
topple *(v.)*	tomber, s'écrouler, chavirer
transform *(v.)*	transformer
transmit *(v.)*	transmettre
trench	tranchée
triangle	triangle
value	valeur
ventilation	ventilation
visual	visuel
walnut	noix
warehouse	entrepôt
weaken *(v.)*	affaiblir
whitewood	bois blanc

Reading Files • File 1
Woodworking

ash	centre
assortment	assortiment
back saw	scie à dos
beam	poutre
beech	hètre
bond	adhésion, lien
brace	attache, vilebrequin à main
carpenter	menuisier
ceiling	plafond
cherry	cerisier
chip	particule
chipboard	aggloméré
chisel	ciseau, burin
clamp	valet (d'établi), serre-joints
cross-head screw	vis à tête cruciforme (Phillips)
durable	durable, endurant
dust	poussière
eccentric grinder	rectifieuse excentrique
edge	bout
even	plat, plan
file	lime
fir	sapin
frame, framework	cadre, armature, bâti, châssis
glazier	vitrier
glue *(n., v.)*	colle ; coller
grease	graisse
grinder	rectifieuse
jigsaw	scie anglaise
joiner	menuisier
keyhole saw	scie à guichet
lath	latte, lame
layman	non initié
liquid	liquide
loose	mal serré
mallet	maillet, mailloche
maple	érable
oak	chêne
pine	pin
planer	raboteur
plywood	contreplaqué
polish *(v.)*	polir
profession	profession
property	propriété
raised-head screw	vis à tête ronde
sand *(v.)*	poncer
shank	tige
shrink *(v.)*	rétrécir, se contracter
sight	vue, viseur
smooth	lisse
solid	solide
solid wood	bois solide
splinter	éclat
star drive screw	vis à six pans creux
substitute	substituer
surface	surface
tacker	cloueuse
thereby	ainsi
thread	pas, filetage
trade	commerce
walnut	noix, noyer
windowpane	vitre
workbench	banc de travail, établi

Reading Files • File 3
Laser technology

absorb *(v.)*	absorber
accurate	précis
acronym	acronyme
amazing	étonnant
amplification	amplification
back and forth	de va-et-vient, de long en large
beam	poutre
behaviour	comportement
billion	billion
bump up *(v.)*	(ici) charger sur
chromium	chrome
coherence	cohérence
crystal	cristal
dental	dentaire
diode	diode
drain *(v.)*	égoutter
drop *(v.)*	laisser tomber
emission	émission
emit *(v.)*	émettre
energy level	niveau d'énergie
entrance	entrée
excess	excès
fascinating	fascinant
flashlight	lampe de poche
focus *(v.)*	focaliser, mettre au point, fixer, centrer, concentrer
give off *(v.)*	dégager
inner level	niveau interne
interact *(v.)*	interagir
laser flash	éclair de laser
light pulse	impulsion de lumière
medium	médium, milieu
mirror	miroir, glace
nucleus	noyau
orbit	orbite
outer level	niveau externe
particle	particule
photon	photon

plastic surgery	chirurgie esthétique
point at *(v.)*	indiquer
radiation	radiation, rayonnement
range	gamme
region	région
release *(v.)*	relâcher
ruby	rubie
surgery	chirurgie
wavelength	longueur d'ondes

Reading Files • File 4 Logistics

accept *(v.)*	accepter
acceptable	acceptable
accomplish *(v.)*	accomplir
air carrier	entreprise de transport aérien
amount	montant, nombre, quantité, somme
arrange for *(v.)*	faire le nécessaire pour
at least	au moins
belly	ventre
border	bordure
cargo plane	avion-cargo
carry out *(v.)*	effectuer
clear *(v.)*	(ici) terminer le chargement
consecutive	consécutif
consume *(v.)*	consommer
consumption	consommation
customer	client
customs	douane
damage	dommages, endommagement
deliver *(v.)*	livrer
(be) dependent on	dépendre de
destination	destination
disaster	désastre, catastrophe
effective	effectif
efficient	efficace
enterprise	entreprise
final leg	dernière étape
fit *(v.)*	être juste, convenir
flawless	parfait
flooding	inondation
freight forwarder	transporteur
freight	fret
graph	graphique
harbour	port
hire *(v.)*	louer
huge	énorme
implement *(v.)*	implémenter
in order to	pour
incredible	incroyable
indispensable	indispensable
insurance	assurance(s)
insure *(v.)*	assurer
involved	concerné
jouney	trajet, voyage
just-in-time	juste-à-temps
labour	main d'œuvre
logistics	logistique
loss	perte
means of transport	moyen de transport
meet the requirements	satisfaire aux exigences
motorway *(UK)*	autoroute
once	une fois, à une époque
origin	origine
originate *(v.)*	être issu de
output	production
prime example	exemple parfait
principal	principal
process	procédure
related	qui s'y rattache
reliability	fiabilité
requirement	exigence, demande
roll-on-roll-off vehicle	transbordeur, manutention par roulage
semitrailer	semi-remorque
shipment	cargaison, envoi, expédition
storage	stockage
suburban	de banlieue
supplier	fournisseur
task	tâche
theft	vol
time consuming	qui prend beaucou de temps
transship *(v.)*	transborder
urban	urbain
waste	déchets
yard	cour, dépôt

Alphabetical wordlist

US	=	American English	**pl.**	=	plural
UK	=	British English	**sing.**	=	singular
v.	=	verb			
n.	=	noun			

A

a large extent	grande mesure
a plus	un plus
about to	sur le point de
abbreviate *(v.)*	abréger
abbreviation	abréviation
above	au-dessus (de)
abrasion	abrasion
abrasive	abrasif
absorb *(v.)*	absorber
abundance	abondance
abundant	abondant
accept *(v.)*	accepter
acceptable	acceptable
access (to)	accès (à)
accessibility	accessibilité
accessory	accessoire
accident	accident
accompany *(v.)*	accompagner
accomplish *(v.)*	accomplir
according to	selon
account	compte
accounting	comptabilité
accuracy	précision
accurate	précis
achieve *(v.)*	réaliser
achievement	exploit, réussite
acid	acide
acoustic insulation	isolement acoustique
acronym	acronyme
addition	addition
addressee	destinataire
adhesive	adhésif
adjust *(v.)*	régler
adjustable wrench	clé à molette
adjustment	réglage
administration	administration
administration building	bâtiment administratif
administrative offices	bureaux administratifs
admission	entrée
advantage	avantage
advertise *(v.)*	faire de la publicité, passer une annonce
advertisement / ad	annonce, (une) publicité
advertising	la publicité
advice	conseil(s)
advise *(v.)*	conseiller
advisor, adviser	conseil, conseiller
aerosol	aérosol
affect *(v.)*	avoir un effet sur
affordable	(dont le prix est) abordable
after-sales service	service après-vente
agree on *(v.)*	convenir de, se mettre d'accord sur
air carrier	entreprise de transport aérien
air conditioning	climatisation
aisle	allée
alert	alerte
aligned	aligné
Allen key(s)/hexagon key	clé mâle
allowance	allocation
alloy	alliage
alter *(v.)*	modifier, changer
altering	modification
alternating current	courant alternatif
alternator	alternateur
altitude	altitude
aluminium *(UK)*, **aluminum** *(US)*	aluminium
amazing	étonnant
ammonium chloride	chlorure d'ammonium
amount	montant, nombre, quantité, somme
amount to *(v.)*	s'élever à
amp	ampère
amplification	amplification
and the like	et d'autres choses comme cela
angle	angle
angle grinder	disqueuse
animated	animé
animation skills	compétences d'animateur
anniversary	anniversaire
announcement	annonce
annoy *(v.)*	embêter
annoying	embêtant
anomaly	anomalie
answering machine	répondeur
anticipate *(v.)*	anticiper
apart from	à part
apartment block	immeuble d'habitation
apologise *(v.)*	s'excuser, regretter
appliance	appareil
applicant	candidat
application	application ; demande d'emploi, lettre de candidature
application software	logiciel d'application
apply *(v.)*	appliquer
apply for *(v.)*	poser sa candidature pour, faire une demande de

appointment	rendez-vous
appreciate *(v.)*	apprécier
apprentice	apprenti
appropriate	approprié, qui convient
approval	approbation
approximately	approximativement
apron	tablier
area code	code région
arrange *(v.)*	organiser
arrange for *(v.)*	faire le nécessaire pour
arrival	arrivée
arrow	flèche
art	art
artificial	artificiel
as far as ... is concerned	en ce qui concerne ...
asbestos	amiante
ash	centre
asparagus	asperge
aspect	aspect
assemble *(v.)*	assembler
assembly line	chaîne de montage
assistance	aide
assistant	assistant, adjoint
assortment	assortiment
assure *(v.)*	assurer
astounding	étonnant
at least	au moins
attached	attaché, joint
attached to	attaché à
attachment	pièce jointe, fichier joint
(for the) attention of	à l'attention de
attraction	attraction
attribute to *(v.)*	attribuer à
audience	public
audit *(v.)*	vérifier, apurer
authorized	autorisé
auto shut-off	arrêt automatique
auxiliary handle	manche auxiliaire
auxiliary scale	échelle auxiliaire, micromètre
availability	disponibilité
available	disponible
average	moyen(ne)
average temperature	température moyenne
avoid *(v.)*	éviter
aware of	conscient de
axle	axe, essieu

B

back and forth	de va-et-vient, de long en large
back saw	scie à dos
backspace	(touche) retour
backup	(de) sauvegarde, (de) secours
baggage reclaim	livraison des bagages
ball bearing	bille de roulement
band saw	scie à ruban
bank account	compte bancaire
bar chart	histogramme
bar graph	histogramme
base	base
base table	table
base unit	unité de base
based on	basé / fondé sur
basic CAD skills	compétences de base en CAO
basic rule	règle de base
basic training	formation de base
battery charger	chargeur de pile / batterie
BC (= before Christ)	avant J.-C.
beam	poutre, rayon
beamer	protecteur
bear *(v.)*	supporter, soutenir
bearings	palier
bed	base, bâti
beech	hètre
beef	bœuf
behaviour	comportement
belly	ventre
belt	courroie
bend *(v.)*	plier
benefit package	les avantages
beverage	boisson
bill *(UK)*	addition
billing	facturation
billion	billion
biro	stylo à bille, bic
bit	foret
blank space	espace
block	bloc
blocked	bloqué
blunt	peu tranchant
board	tableau
boldface	gras
bolt	boulon
bond	adhésion, lien
bond *(v.)*	(s')adhérer, lier, unir
bonus	bonus, prime
booked up	complet
bookend	serre-livres
boot error	erreur d'initialisation
booth	cabine
border	bordure
border	bordure
bore *(v.)*	percer
bottom	fond
brace	attache, vilebrequin à main
brad point	pointe à tête perdu
branch	branche, agence

brass	laiton
brick	brique
brick course	assise
brief	bref
brief outline	idée générale
brittle	cassant, fragile
brittleness	fragilité
brochures	brochures, documentation
bronze	bronze
brook	petite rivière
brush	brosse
bucket	seau
buckle *(v.)*	boucler
bulb	ampoule
bulk buyers	acheteurs en gros
bump up *(v.)*	(ici) charger sur
business relation	relation d'affaires
(get/hear the) busy signal	sonner occupé
button	bouton, touche
buttress	contrefort
by return / by return mail	par retour de courrier

C

calculate *(v.)*	calculer
calculating machine	machine à calculer
calculator	calculette
calibrate *(v.)*	calibrer
calm *(v.)*	calmer
Can I take a message?	Puis-je prendre un message ?
Can you manage alone?	Pouvez-vous vous débrouiller tout seul ?
cancer	cancer
capable	capable
capacitor	condensateur
capacity	capacité
capital letter	majuscule
capitalisation	capitalisation
caps lock	majuscule verrouillée
car component supplier	fournisseur de composants pour l'industrie automobile
car park *(UK)*	parking
car rental	location de voiture
carbon	carbone
carbon dioxide	dioxyde de carbone
care	soin
career prospect	perspective de carrière
cargo plane	avion-cargo
carpenter	menuisier
carry out *(v.)*	effectuer
cartridge	cartouche
casing	revêtement
cast *(v.)*	couler, fondre
cast iron	fonte
catalytic converter	pot catalytique
cattle	bétail
ceiling	plafond
celebrate *(v.)*	fêter
cell(ular) phone *(US)*	téléphone portable
cement	ciment
central processing unit (CPU)	unité centrale de traitement (UCT)
centre lathe	tour à pointes
centre-punch	pointeau, amorçoir
ceramic(s)	céramique
certainly	certainement
CFC gases	gaz CFC
chainsaw	tronçonneuse
challenge	challenge, défi
challenging environment	environnement stimulant
chamber of commerce	chambre de commerce
character	caractère
charge *(v.)*	charger
charger	chargeur
charges	charges
chart	schéma, diagramme
charter	charte
check *(US)*	addition
check *(v.)*	vérifier
check the account	vérifier le compte
check-in	enregistrement, arrivée, accueil
check-out	heure de départ (hôtel)
chef	chef (de cuisine)
cherry	cerisier
chest of drawers	commode
chief executive	directeur général
chilli	chili
chimney stack	cheminée
chip	particule
chipboard	aggloméré
chisel	ciseau, burin
chlorofluorocarbon	chlorofluorocarbone
choice	choix
chromium	chrome
chuck	mandrin
chutney	chutney (condiment à base de fruits)
circle	cercle
circular	circulaire, rond
circumstance	circonstance
civil engineer	ingénieur des travaux publics
civil engineering	génie civil
claim	réclamation
clamp	valet (d'établi), serre-joints
clamp *(v.)*	serrer, cramponner

clamp screw	vis d'arrêt/de blocage
clamping jaw	mordache
clamping system	système de blocage
clay	argile
clear *(v.)*	(ici) terminer le chargement
client	client
clip	collier, bague, pince
clipboard	calepin
clockwise	dans le sens des aiguilles d'une montre
clutch	embrayage
coherence	cohérence
cohesion	cohésion
coil	bobine
colleague	collègue
colonial age	époque coloniale
Columbus Day *(US)*	jour commémorant l'arrivée de Christophe Colomb en Amérique (2e lundi d'octobre)
column	colonne
column drill	perceuse à colonne, fraiseuse
combination pliers	pince universelle
combine *(v.)*	combiner
command	commande
commemorative	commémoratif
common	fréquent, commun
commutator	commutateur
compact (type of car)	petite voiture
compact *(v.)*	compacter
compartment	compartiment
competence	compétence
complaint	réclamation
complemented	complété, accompagné
complete *(v.)*	compléter, achever
complimentary	gratuit, gracieux
complimentary close	salutation
component	composant
composite	composite
compound	composé
compressive strength	force
Computerized Numerical Control (CNC)	commande numérique
concerned	concerné
(be) concerned with	concerner
concerning	concernant
conclusion	conclusion
concrete	béton
concrete base	base / soutien en béton
concrete block	bloc de béton
conduct *(v.)*	conduire
conductive	conducteur
conductivity	conductivité
confident	confiant
confirm *(v.)*	confirmer
confuse *(v.)*	confondre
connect *(v.)*	connecter
connected in parallel	connecté en parallèle
connected in series	connecté en série
connecting rooms	chambres communicantes
connection	lien
conquer *(v.)*	conquérir, vaincre
consecutive	consécutif
consequently	par conséquent
consist of *(v.)*	consister en
consume *(v.)*	consommer
consumer report	sondage des consommateurs
consumption	consommation
continual	continu
contract	contrat
control *(n., v.)*	contrôle ; contrôler, régler
control panel	panneau de contrôle
control panel	tableau de bord
controls	contrôles
controversial issue	sujet de controverse
convenience	confort, commodité
convenient	bien situé, commode
coolant	caloporteur, liquide de refroidissement
cope with *(v.)*	s'occuper de
copper	cuivre
cord	cordon, câble
cordial	cordial
cordless	sans fil
cornmeal	farine de maïs
correct *(v.)*	corriger
correspondence	correspondance
corresponding	correspondant
corrode *(v.)*	corroder
corrosion	corrosion
cost accountant	comptable spécialisé en comptabilité analytique
cost-intensive country	pays où les coûts de production sont élevés
counter- / anti-clockwise	en sens inverse des aiguilles d'une montre
countersunk screws	vis noyées
country code	code pays
courtesy	gratuit, service offert aux clients
crab cake	croquette de crabe
crack	fissure
cracked	fendu
craftsman	artisan
cranberry	airelle
crash	incident
cream	crème

create *(v.)*	créer
crib	lit d'enfant
crisp	croquant
cross-head/cross-slot screw	vis à tête cruciforme (Phillips)
cross-slide	chariot longitudinal
crossbar	barre de traverse / d'accouplement
crown	couronne
crystal	cristal
crystalline	cristallin
cube	cube
cucumber	concombre
current	courant
current rates	taux en vigueur
current surge preventing	empêcher les à-coups de courant
currently	actuellement
cursor	curseur
curved	courbé
customer	client
customer service	service clientèle
customs	douane
cut-and-paste	couper-coller
cutting edge	coupant, tranchant
cutting tool	outil de découpage / cisaillage
CV (Curriculum Vitae)	CV
cylinder	cylindre

D

damage *(n., v.)*	dommages, endommagement ; endommager
damp	humide
database	base de données
DC (direct current)	courant continu
dead load	poids mort
deadline	temps limite, délai
debate	débat
debur *(v.)*	ébarber, ébavurer
decade	décennie
decide on *(v.)*	se décider pour
decision	décision
deep	profond
default settings	réglage / paramétrage par défaut
defect	défaut
defence system	système de défense
deflection	déflection
deform *(v.)*	déformer
(Honours) degree	licence
deinstallation	désinstallation
delay	retard
delete *(v.)*	effacer
deliver *(v.)*	livrer
delivery	livraison
demand	demande
demonstrate *(v.)*	démontrer, présenter
dental	dentaire
dentist	dentiste
depart *(v.)*	partir
Department for Employment and Education *(UK)*	ministère britannique de l'Emploi et de l'Éducation
departure	départ
depend on *(v.)*	dépendre de
dependent child	enfant à charge
(be) dependent on	dépendre de
depth gauge	calibre de profondeur
derive *(v.)*	provenir, venir
derogatory	dérogatoire
design	conception, dessin
design *(n., v.)*	plan, conception, dessin ; créer, concevoir, dessiner
desk	bureau
desk tidy	organiseur de bureau
desktop organiser / organizer	agenda électronique de bureau
dessert	dessert
destination	destination
destroy *(v.)*	détruire
detergent	détergent
determine *(v.)*	déterminer
detrimental	nuisible, préjudiciable
deuterium	deutérium
develop *(v.)*	(se) développer
development	développement, évolution
device	appareil
diagram	diagramme
dial *(v.)*	composer (un numéro)
dialogue box	boîte de dialogue
diameter	diamètre
dictionary	dictionnaire
dimension	dimension
diode	diode
direct current	courant continu
directory enquiries	renseignements
disadvantage	inconvénient
disaster	désastre, catastrophe
disastrous	désastreux
discharge *(v.)*	décharger
discharged	déchargé
discharging process	processus de déchargement
discount	remise
discount store	solderie
disease	maladie
display *(v.)*	afficher
disposal	enlèvement, traitement

dispose of *(v.)*	se débarasser de
dissolve *(v.)*	dissoudre
distance	distance
distort *(v.)*	distordre
distribution	diffusion, distribution
divide *(v.)*	diviser
divorced	divorcé
do a favour *(v.)*	rendre un service
do the filing *(v.)*	faire le classement
Do-It-Yourself (DIY)	bricolage
domain	domaine
domestic	domestique, national
domestic flight	vol intérieur
door knob	poignée de porte
door lock	serrure de porte
dot	point
double insulation	isolement double
double room	chambre double
down the hall	à l'autre bout du couloir
download *(v.)*	télécharger
downtown *(US)*	centre ville
draft / draught beer	bière pression
drag and drop	glisser et déposer
drain *(v.)*	égoutter
drainage	système d'égouts / d'évacuation des eaux
(electric) drill	perceuse (électrique)
drill *(v.)*	percer
drill bit	foret de perceuse
drill the surface	percer la surface
drive in *(v.)*	enfoncer
driver's licence	permis de conduire
drop *(v.)*	laisser tomber
drop off *(v.)*	laisser, déposer
drown *(v.)*	noyer
dry cleaning	pressing
duckling	caneton
ductility	ductilité, malléabilité
due to	à cause de
durability	durabilité
durable	durable, endurant, résistant
dust	poussière

E

e.g. (= for example)	par exemple
eccentric grinder	rectifieuse excentrique
ecological	écologique
economy class	classe touriste
edge	bout
edit *(v.)*	éditer
effective	effectif
efficiency	efficacité
efficient	efficace
either ... or	ou ... ou
elasticity	élasticité
electric charge	charge électrique
electrical engineer	ingénieur électricien
electrolyte	électrolyte
electrostatic induction	induction électrostatique
ellipse	ellipse
emergency	urgence
emergency escape route	sortie de secours
emergency sign	panneau
emergency stop	arrêt d'urgence
emission	émission
emit *(v.)*	émettre
emphasis	accent
employ *(v.)*	employer, utiliser
employee	employé, salarié
employer	employeur
employment opportunity	perspective d'emploi
enable *(v.)*	permettre, valider
enclosure	pièce jointe
enduring	durable
energy consumption	consommation d'énergie
energy level	niveau d'énergie
engaged	occupé
engine	moteur
engine test facility	installation pour tester les moteurs
engineering	génie
engineering degree	diplôme / licence en ingénierie
enquiry	demande de renseignements
ensure *(v.)*	assurer
enter *(v.)*	entrer
enterprise	entreprise
entire	entier
entrance	entrée
entrée *(US)*	plat principal
entry	entrée
envelope	enveloppe
environment	environnement
environmentally-friendly	respectueux de l'environnement
equal *(v.)*	égaler
equipment	équipement, matériel
equipped with	équipé de
equivalent	équivalent
escape route	voie de secours, sortie de secours
escape	s'échapper
especially	particulièrement, spécialement
essential	essentiel
establish *(v.)*	établir
estimate *(v.)*	estimer, faire une estimation

estuary	estuaire
even	plat, plan, régulier, uniforme
exaggerate *(v.)*	exagérer
excavate *(v.)*	excaver, creuser
exceed *(v.)*	dépasser
excess	excès, surplus
exercise gym	salle de gym
exhaust pipe	pot d'échappement
exhibit *(v.)*	exposer
exit	sortie
expand *(v.)*	agrandir, développer
expect *(v.)*	s'attendre à
expectation	attente
experience *(n., v.)*	expérience ; expérimenter, vivre
explain *(v.)*	expliquer
explanation	explication
exploded view	vue éclatée
express	express
extend *(v.)*	(s')étendre, (se) développer
extension	poste
extension number	numéro de poste
extensive	important, considérable
external	externe
external measurement	mesure / dimension extérieure
extractor	extracteur, ventilateur

F

face *(v.)*	être en face de
face down	face contre la vitre (photocopies)
facilitate *(v.)*	faciliter
facilities	installations, équipements
facility	facilité
failure	échec
fair	foire, salon
fair *(n.)*	foire, salon
fairly	assez
(be) familiar with	bien connaître
family name	nom de famille
fascinating	fascinant
fasten *(v.)*	attacher
fatal	mortel
faulty	défectueux
favourable	favorable
feature	caractéristique, particularité
feed *(v.)*	alimenter
feeder road	route d'accès
feeler gauge	jauge d'épaisseur de profondeur / de palpage
female	féminin
ferrous metals	métaux ferreux
fertilizer	engrais
fibreglass	fibre de verre
figure	figure
figure out *(v.)*	arriver à comprendre
file *(n., v.)*	dossier, fichier, lime ; limer
film speed	vitesse de film
final leg	dernière étape
finalize *(v.)*	finaliser
finding	résultat, conclusion
finish	fin, terminaison
finished goods	produits finis
fir	sapin
fire extinguisher	extincteur
fire sign	panneau : risque d'incendie
first aid kit	trousse à pharmacie
first course	entrée
first name	prénom
fit *(v.)*	être juste, convenir
fixed	fixé
fixed jaw	mâchoire fixe
flashlight	lampe de poche
flawless	parfait
flicker-free	sans scintillement
flight attendant	steward, hôtesse de l'air
float	flotte
flooding	inondation
floppy disk	disquette
foam pillows	oreillers en mousse
foamed	écumé
focus *(v.)*	focaliser, mettre au point, fixer, centrer, concentrer
folding rule	mètre pliant
(be) fond of	aimer
font size	taille de police
fonts	polices de caractères
footer	titre de bas de page
for instance	par exemple
forbid *(v.)*	interdire
fork lift truck	chariot à fourche, gerbeuse
format	format
format *(v.)*	formater
fossil fuels	combustibles fossiles
foundation	fondation
frame, framework	cadre, armature, bâti, châssis
freelancer	travailleur indépendant
freight	fret
freight forwarder	transporteur
frequency	fréquence
frequent	fréquent
friction	friction

fried	frit, poêlé
fringe benefit(s)	avantage(s) en nature
from conception to completion	de la conception à la réalisation
fuel cell	pile à combustible
full-size (type of car)	grosse voiture
fumes	exhalaisons, fumées
function space	salle de réception
funds	fonds
furnace	fourneau
furniture	mobilier
fuse	fusible
fuse together *(v.)*	fusionner

G

gadget	gadget
gain a foothold *(v.)*	prendre pied
game	gibier
gangway	passage
gap	écart
gate	porte, portail
gather *(v.)*	cueiller, recueillir
gear	engrenage, vitesse
gear and control unit	commande de variateur de vitesses
gearbox	boîte de vitesses
general	général
general workshop	atelier
generate *(v.)*	générer
genius	génie
get injured	être blessé
giant	géant
give off *(v.)*	dégager
give sombody a ring	passer un coup de fil à quelqu'un
glacier	glacier
glazed	glacé
glazier	vitrier
global	global, mondial
glove	gant
glue *(n., v.)*	colle ; coller
goggles	lunettes de travail
goldsmith	orfèvre
goods	biens, marchandises
graduate *(n., v)*	diplômé ; obtenir son diplôme / sa licence
grammar checker	vérificateur de grammaire
grant *(v.)*	accorder
granulate	granulé
graph	graphique
graphic tool	outil graphique
grease	graisse
grinder	rectifieuse, meuleuse, pileur, broyeur
grinding machine	rectifieuse, affûteuse
grinding wheel	disque abrasif
guarantee *(v.)*	garantie
guard	dispositif de protection
guess *(v.)*	deviner
guide *(v.)*	guider
guideline(s)	ligne directrice, directives

H

hacksaw	scie à métaux
hairdryer	sèche-cheveux
hand luggage	bagage à main
handle *(n., v.)*	manche, manette ; manier, traiter
handout	prospectus, polycopié
handshield	protège-main
hang on *(informal)*	ne quittez pas
harbour	port
hard	dur
hard disk	disque dur
harden *(v.)*	durcir
hardly	à peine, ne ... guère
hardness	dureté
hardware	machines
harmful	nuisible, nocif
harmful to the health	nuisible à la santé
harness *(v.)*	harnais
have ... in common	avoir ... en commun
header	en-tête
headstock	poupée (d'un tour)
height	hauteur, taille
help desk	hotline, assistance en ligne
herbed	avec des herbes
herbs	herbes
hero	héros
heroine	héroïne
hertz (Hz)	hertz
hesitate *(v.)*	hésiter
high speed steel (HSS)	HSS
high tide	marée haute
high-rise building	tour
highly skilled	très qualifié
hint	allusion, indice
hire *(v.)*	louer
historic square mile	centre historique
hold on *(formal)*	ne quittez pas
hold the line	ne pas quitter
hole	trou
hole punch	perforateur
hollow	creux
Home Office *(UK)*	ministère britannique de l'Intérieur
horseradish	raifort
house *(v.)*	recevoir, contenir

housing	capot
however	pourtant
huge	énorme
human resources adviser	conseil en ressources humaines
hydraulics	hydraulique
hydroelectric power	énergie hydroélectrique
hydrogen	hydrogène
hypertext	hypertexte
hyphenation	coupure de mot

I

I didn't catch that.	Je n'ai pas bien compris.
I know my way round now.	Je connais bien les locaux maintenant.
I wonder if …	Je me demande si …
I'll be right with you.	J'arrive.
icon	icône
image	image
imagine *(v.)*	imaginer
immense	immense
immerse *(v.)*	immerger
immigration	immigration
immigration card	carte d'immigration
impact drill	perceuse à percussion
impact drilling	perçage à percussion
imperial measure (imperial system of measurement)	mesure britannique
impetus	élan, force d'impulsion
implement *(v.)*	implémenter
implementation	implémentation
impose *(v.)*	imposer
imposed load	charge imposée
impression	impression
improper	inadéquat, mauvais
improve *(v.)*	(s')améliorer
improvement	amélioration
in addition	en plus
in charge of	responsable de
in order to	pour
in public	en public
in spite of	malgré
in the long term	à long terme
inappropriate	inapproprié
include *(v.)*	inclure
incorporate *(v.)*	incorporer
increase *(v.)*	augmenter
incredible	incroyable
indefinitely	indéfiniment
Independence Day *(US)*	fête nationale de l'Indépendance (le 4 juillet)
indicate *(v.)*	indiquer
indispensable	indispensable
information management	gestion d'information
information sheet	feuille d'information
ingredient	ingrédient
initial research	recherche initiale
injured	blessé
ink jet (printer)	(imprimante) à jet d'encre
inlay	incrustation
inn	auberge
inner level	niveau interne
inorganic	inorganique
input device	périphérique d'entrée
inscribe *(v.)*	inscrire
insert *(v.)*	insérer
inside measurement	mesure / dimension intérieure
insist on *(v.)*	insister sur
inspection	inspection
instantly	instantanément
instead	à la place
instead of	au lieu de
instruction	instruction
insulation	isolement
insurance	assurance(s)
insure *(v.)*	assurer
integrate *(v.)*	intégrer
integrated circuit	circuit intégré
intend *(v.)*	avoir l'intention de
intensify *(v.)*	intensifier
interact *(v.)*	interagir
interconnection	interconnection
interest rate	taux d'intérêt
intermediate	intermédiaire
internal combustion engine (ICE)	moteur à combustion interne
internal measurement	mesure / dimension intérieure
internal operation	opération interne
interrupt *(v.)*	interrompre
introduce *(v.)*	présenter, introduire
introduction	introduction
invent *(v.)*	inventer
inventor	inventeur
inverter	inverteur
invoice	facture
involve *(v.)*	comprendre, consister en
involved	concerné
iron	fer, fer à repasser
ironing board	planche à repasser
isolated	isolé
issue *(v.)*	émettre
item	article

J

jacuzzi	jacuzzi
jammed	être bloqué

jaw	mâchoire
jewellery	bijoux
jigsaw	scie anglaise
job vacancy	poste à pourvoir
join *(v.)*	joindre, attacher
joiner	menuisier
joint	joint
joist	solive
jouney	trajet, voyage
joystick	manette de jeu
jumbo	géant
jumper	pullover
just-in-time	juste-à-temps

K

keep clear *(v.)*	(se) mettre à l'écart
keep track of *(v.)*	suivre
key	clé, touche
keyboard	clavier
keyhole saw	scie à guichet
keyless chuck	mandrin sans clé
keyword	mot clé
kinetic energy	énergie kinétique
knee	capot
knowledge	connaissances

L

label *(n., v.)*	étiquette ; étiqueter
Labo(u)r Day *(US)*	fête du travail (1er lundi de septembre)
labour	main d'œuvre
lamb	agneau
lamp socket	douille
landscape	paysage
laptop	portable
larch	mélèze
large	grand
large-scale (production)	à grande échelle
large-scale manufacturing	fabrication à grande échelle
laser flash	éclair de laser
lateral	latéral
latest	dernier
lath	latte, lame
lathe	tour
laundry	buanderie
lawyer	avocat
layman	non initié
layout	maquette, disposition
lead	extension, rallonge ; plomb
leaf through *(v.)*	feuilleter
leaflet	dépliant
leather	cuir
legal	juridique, légal
lemon	citron
lemongrass	citronnelle
length	longueur
lens	lentille, objectif, verre
lesson	leçon
letter of complaint	lettre de réclamation
lever	levier
light pulse	impulsion de lumière
light-duty	léger
lightning	éclair
lime	chaux
limitless	sans limites
line graph	courbe
link *(v.)*	lier, connecter, chaîner
link together *(v.)*	lier, joindre
liquid	liquide
live conductor	conducteur sous tension
live load	poids roulant
live wire	fil sous tension
load *(n., v.)*	charge ; charger
load-bearing	porteur
load-bearing wall	mur porteur
lobster	homard
local	local
locate *(v.)*	localiser
located	situé
lock out	blocage
locked in	(ici) engagé
locking lever	levier de blocage
logistics	logistique
longitudinal	longitudinal
loose	mal serré
loosen *(v.)*	déserrer
loss	perte
low tide	marée basse
low-cost country	pays où les coûts de production sont bas
lower *(v.)*	baisser
lubricant	lubrifiant
lubricate *(v.)*	lubrifier
lubricating oil	huile de graissage
lubrication	lubrification

M

machine tool(s)	machine-outil(s)
machining	fabrication à la machine, usinage
main course	plat principal
main entrance	entrée principale
main scale	échelle principale (en millimètres)
main switch	interrupteur principal (marche-arrêt)

mainly	principalement
mains	secteur
mains testing screwdriver	tournevis testeur
maintenance	maintenance
major	majeur, important
major benefit	bénéfice principal
male	masculin
malleability	malléabilité
mallet	maillet, mailloche
managing director	directeur général
mandatory sign	panneau obligatoire
manganese	manganèse
manipulate *(v.)*	manipuler
manual	manuel
manually	manuellement
manufacture *(n., v.)*	fabrication ; fabriquer
manufacturer	producteur, fabricant
manufacturing	fabrication
manufacturing engineer	ingénieur de la fabrication
manufacturing unit	unité de fabrication
maple	érable
marble	marbre
marital status	état civil
mark *(v.)*	marquer
market research	étude de marché
marketing manager	responsable du marketing
masonry	maçonnerie
master	maître
materials in stock	matériaux en stock
mean temperature	température moyenne
means *(n. sing.)*	moyen
means of transport	moyen de transport
measure *(n., v.)*	mesure ; mesurer
measurement	mesure
mechanical engineering	mécanique
medium *(adj.)*	à point (steak)
medium	médium, milieu
meet (needs) *(v.)*	répondre à (des besoins)
meet human requirements	satisfaire aux exigences humaines
meet someone's needs	répondre aux besoins de quelqu'un
meet the requirements	satisfaire aux exigences
melt *(v.)*	(faire) fondre
melting point	point de fusion
membrane keyboard	clavier avec bâche de protection
membrane switch	interrupteur avec bâche de protection
Memorial Day (US)	jour férié en l'honneur des soldats américains morts pour la patrie (dernier lundi de mai)
memory	mémoire
mention *(v.)*	mentionner, faire mention de, citer
menu	carte
merge *(v.)*	fusionner
message	message
metal bar	barre de métal
methane	méthane
micrometer	micromètre
mid-size (type of car)	voiture de taille moyenne
middle name	deuxième prénom
mild steel	acier doux
milling machine	fraiseuse
minor	mineur
minor thing	chose mineure
mirror	miroir, glace
mishandling	mauvais usage
misspelling	faute d'orthographe, coquille
misunderstanding	malentendu
misuse	mauvais usage, usage abusif
mobile phone *(UK)*	téléphone portable
modification	modification
monitor	moniteur
more likely	plus probable
motion	mouvement
motor-reversing switch	interrupteur pour inverser le moteur
motorway *(UK)*	autoroute
mould *(v.)*	mouler
mount *(v.)*	monter, installer
mousepad	tapis de souris
mouth-to-mouth	bouche-à-bouche
mov(e)able	mobile
mud guard	garde-boue
multi-purpose	à usages multiples, polyvalent
mushroom	champignon
mustard	moutarde

N

nail *(n., v.)*	clou ; clouer
narrow	étroit
national	national
neat	net, soigné, bien rangé
negotiable	à négocier
network support technician	technicien de maintenance de réseau
newsagent	marchand de journaux
newsstand	kiosque à journaux
nitrogen	azote
nitrous oxide	oxyde nitreux
noise	bruit

non-allergenic	hypoallergénique
nonius	micromètre
notepad	cahier
nowadays	de nos jours
nuclear fission / fusion	fission / fusion nucléaire
nuclear waste	déchets nucléaires
nucleus	noyau
(be a) nuisance	être énervant
numerous	nombreux
nut	écrou

O

oak	chêne
obligatory	obligatoire
obtain *(v.)*	obtenir
occasionally	de temps en temps
occur *(v.)*	se passer
octagonal	octagonal
of course	bien sûr
offer *(n., v.)*	offre ; offrir
official	officiel, agent
old-fashioned	démodé
on file	sur fichier
once	une fois, à une époque
operate *(v.)*	opérer
operate a machine *(v.)*	faire fonctionner une machine
optimal solution	solution optimale
option	option
orbit	orbite
order *(n., v.)*	commande, ordre ; commander, ordonner
ore	minerai
organigram	organigramme
origin	origine
originate *(v.)*	être issu de
outdated	démodé
outer level	niveau externe
output	production
output device	périphérique de sortie
outside measurement	mesure / dimension externe
oval	ovale
overarm	traînard
overheating	échauffement
overtime	heures supplémentaires
overview	vue d'ensemble
overwrite *(v.)*	surfrapper
oxide	oxyde
oyster	huître

P

packaging	conditionnement
pancake	crèpe
panel discussion	débat, tribune
paper clip	trombone
paper jam	bourrage papier
paraffin	pétrole, mazout
paragraph	paragraphe
paramount	primordial
parking lot *(US)*	parking
particle	particule
particular specialization	spécialisation particulière
partition (wall)	cloison
pass *(v.)*	passer
passenger	passager
patch	modification de programme
path	chemin
pattern	modèle, motif, système
peaceful	paisible, tranquille
peach	pêche
peak hour	heure de pointe
pecan	pecan
pellet	boulette
pencil sharpener	taille-crayon
penetration	pénétration
pepper	poivre
percentage	pourcentage
percussion drilling	perçage à percussion
perform *(v.)*	effectuer, opérer, accomplir
performance	fonctionnement, performance
period *(US)*	point
peripherals	périphériques
permanent	permanent
personnel	personnel
pewter	étain
phosphorus	phosphore
photon	photon
photovoltaics	photovoltaïque
pick someone up *(v.)*	aller chercher quelqu'un (en voiture)
pick up *(v.)*	ramasser
pickapeppa remoulade *(US)*	mayonnaise à la paprika
pie chart	camembert
pier	pile
pierce *(v.)*	percer
pigeonhole	casier (à courrier)
pile	pile, tas
pilot hole	avant-trou
pine	pin
pitched roof	toit en pente
planer	raboteur
plant	plante, usine
plastic surgery	chirurgie esthétique
plastics	plastiques
platinum	platine

plentiful	abondant
plotter	table traçante
plug	fiche, prise
plumber	plombier
plywood	contreplaqué
pocket calculator	calculette
point at *(v.)*	indiquer
poisonous	vénéneux, toxique
polish *(v.)*	polir
polite	poli, bien élevé
pollutant	polluant
pollution	pollution
polyamide	polyamide
polygon	polygone
pool	piscine
popular	populaire
porcelain	porcelaine
portable	portable, portatif
possible	possible
postal service	service des postes
pottery	poterie, céramiques
poultry	volaille
power tool(s)	outil électrique
PR (public relations)	relations publiques
pre-cast concrete floor	sol en béton préfabriqué
pre-drill *(v.)*	pré-percer
pre-stressed concrete	béton préconstraint
precautions	précautions
precision	précision
prefer *(v.)*	préférer
preferable	préférable
preferably	de préférence
prepare *(v.)*	préparer
prepared	préparé / prêt
prepay *(v.)*	prépayer
preselection	présélection
present *(v.)*	présenter
presentation	présentation
pressing	pressing
pressure	pression
prevent *(v.)*	empêcher
primarily	principalement
prime example	exemple parfait
prime rib	côte de bœuf
principal	principal
privacy	intimité
privilege	privilège
procedure	procédure
process	procédure
processing	traitement
product range	gamme de produits
production output	production
profession	profession
prohibit *(v.)*	interdire
prohibition	interdiction
project *(v.)*	projeter
promote *(v.)*	promouvoir
(be) promoted	être promu, avoir une promotion
prompt	rapide, prompt
pronounce *(v.)*	prononcer
pronunciation	prononciation
proof	preuve(s)
proofreading	correction d'épreuves
propellant	propulsant, carburant
properties	propriétés
property	propriété
propose *(v.)*	proposer
propulsion	propulsion
protect *(v.)*	protéger
protective	de protection
prove *(v.)*	prouver
provide *(v.)*	fournir
provision	disposition, provision, réserve
public relations manager	responsable des relations publiques
puff pastry	pâte feuilletée
purchase *(n., v.)*	achat ; acheter
purchasing department	service achats
pure	pur
purpose	but, objectif
push sideways *(v.)*	pousser de côté

Q

quality assurance	garantie de qualité
quality engineer	ingénieur de contrôle de qualité
quote *(v.)*	annoncer, indiquer, citer

R

rack of lamb	carré d'agneau
radiation	radiation, rayonnement
radius	rayon
raised-head screw	vis à tête ronde
range	gamme
range of accessories	gamme d'accessoires
rare	saignant (steak)
rarely	rarement
rate	taux, prix
ratio	taux
raw material	matière première
re-mould *(v.)*	remouler, rechaper
reason	raison
recast *(v.)*	refondre
receipt	reçu
receive *(v.)*	recevoir
recent	récent
receptionist	réceptionniste

recharge *(v.)*	recharger
rechargeable	rechargeable
recharging process	processus de rechargement
recognise *(v.)*	reconnaître
recommend *(v.)*	recommander
recommendation	recommandation
recruitment needs	besoins de personnel
rectangle	rectangle
rectangular	rectangulaire
recyclable	recyclable
reduce *(v.)*	réduire
reduction	réduction
redwood	séquoia
refer to *(v.)*	faire référence à, se référer à
reference	référence
refined	raffiné
reflector	réflecteur
refrigerator	réfrigérateur
regard *(v.)*	considérer
regarding	concernant
regardless	quand même
region	région
registration	inscription
reinforced concrete	béton renforcé
reinstall *(v.)*	réinstaller
reject *(v.)*	rejeter
related	qui s'y rattache
relax *(v.)*	se détendre
release *(v.)*	relâcher
relevant	pertinent
reliability	fiabilité
reliable	fiable
rely on *(v.)*	compter sur, faire confiance à
remedy	remède
render *(v.)*	rendre
renewable energy	énergie renouvelable
repeat *(v.)*	répéter
repel *(v.)*	repousser
replace *(v.)*	remplacer
replacement	remplacement
report	rapport
represent *(v.)*	représenter
representative	représentant
reprint *(v.)*	réimprimer
repulsive force	force répulsive
request *(v.)*	demander
require *(v.)*	demander, exiger, requérir
required	demandé, exigé, requis
requirement	demande, exigence
research	recherche(s)
research and development	recherche et développement
research engineer	ingénieur chercheur
reservoir	réservoir
reset *(v.)*	réenclencher, réinitialiser
reshape *(v.)*	reformer
residence permit	permis de séjour
residential	résidentiel
resist *(v.)*	résister à
resistance	résistance
resistant to	résistant à
resistor	résistance
resources	ressources
respective	respectif
respiration	respiration
respiratory failure	défaillance respiratoire
responsibility	responsabilité
restart *(v.)*	recommencer, redémarrer
restrict *(v.)*	restreindre, limiter
result	résultat
retailer	détaillant
retrieve data	récupérer des données
return time	heure de retour
retype *(v.)*	retaper, resaisir
reuse *(v.)*	réutiliser
reverse *(v.)*	inverser
reversible	reversible
reversing lever	levier d'inversion de marche
review *(v.)*	examiner, revoir
revive *(v.)*	raviver
revolution speed	vitesse de révolution
revolutionary	révolutionnaire
revolutionise *(v.)*	révolutionner
right angle	angle droit
right of residence	droit de séjour
rigid	rigide
river control	contrôle fluvial
rivet *(n., v.)*	rivet ; riveter
riveting tongs	pince à rivets
roasted	rôti, cuit au four
rod	barre
roll-on-roll-off vehicle	transbordeur, manutention par roulage
rotate *(v.)*	tourner, (faire) pivoter
rotor	rotor
roughly	approximativement
route	itinéraire, chemin
row	rangée
rpm (revolutions per minute)	tr/min (tours à la minute)
ruby	rubie
rust *(v.)*	rouille

S

saddle	selle
safety	sécurité

safety boots	chaussures de sécurité
safety sign	panneau de sécurité
salutation	salutation
sample	échantillon
sand *(v.)*	poncer
sander	ponceuse
sanitation	système sanitaire, hygiène publique
sauce	sauce
save *(v.)*	sauver, sauvegarder
save as *(v.)*	enregistrer sous
sawdust	sciure
scale	échelle
schedule *(n., v.)*	planning, horaire ; planifier
(natural) science	science (naturelle)
scientist	scientifique
screech *(v.)*	hurler, grincer
screen	écran
screw *(n., v.)*	vis ; visser
screwdriver	tournevis
sea level	niveau de la mer
search *(v.)*	rechercher
search engine	moteur de recherche
seasonal	de la saison, saisonnier
secondary school	lycée
security check	contrôle de sécurité
select *(v.)*	sélectionner
self-employed	indépendant, qui travaille à son compte
self-employed person	personne qui travaille à son compte
semiconductor	semiconducteur
semitrailer	semi-remorque
sense of rotation	sens de rotation
serious	sérieux, grave
serve *(v.)*	servir
server	serveur
service	service
settle *(v.)*	résoudre (un problème)
sew *(v.)*	coudre
shaft	arbre, manche, tige
shallot	échalote
shank	tige
shape	forme
shareholder	actionnaire
sharp rise	forte augmentation, montée en flèche
sharpen *(v.)*	aiguiser
sheet material	matériel en feuilles
sheet metal	métal en feuilles, tôle
sheet steel	tôle d'acier
shiny	brillant
shipment	cargaison, envoi, expédition
shoe shine	cirage des chaussures
shop	(ici) atelier
shop assistant	vendeur
shop floor	atelier
short-term contract	contrat à durée déterminée
shower	douche
shrink *(v.)*	rétrécir, se contracter
shuttering	coffrage
shuttle	navette
sight	vue, viseur
sign *(n., v.)*	panneau ; signer
signature	signature
silicic acid	acide silicique
silicon	silicium
silicon wafer	tranche de silicium
single room	chambre simple
sink	évier
sintered metal	métal fritté
site	site
size	taille
skilled	qualifié
skilled worker	ouvrier qualifié
slideway	coulisse
sliding caliper	calibre à coulisse
slight	léger
slot	fente
slot screw	vis à tête fendue
smoked	fumé
smooth	lisse
soccer club	club de football
socket	godet, cosse, emboîtement
(hexagon) socket screw	vis à tête hexagonale
soften *(v.)*	ramollir, adoucir, assouplir
softness	douceur, moelleux
soil	sol
solar furnace	four solaire
solar panel	panneau solaire
solder *(v.)*	souder
soldering	soudure
soldering iron	fer à souder
solid	solide
solid wood	bois solide
sophisticated	complexe, perfectionné
soup	soupe, potage
source	source
Space bar	Barre espace
span	portée, largeur
spanner	clé
spare part	pièce de rechange, pièce détachée
spark plug	bougie
speak up *(v.)*	parler franchement, parler plus fort
special-interest magazine	revue spécialisée
specific	précis

speed limit	limitation de vitesse
speed-selection lever	levier pour sélectionner la vitesse
spell *(v.)*	épeler
spell checker	vérificateur d'orthographe
spelling	orthographe
spice *(v.)*	épicer
spicy	épicé
spin *(v.)*	(faire) tourner / pivoter
spinach	épinards
spindle	arbre, broche, axe
spiral drill bit	foret en spirale
spirits	spiritueux
splinter	éclat
split *(v.)*	fendre
spouse	époux, épouse
spread *(n., v.)*	étendue, développement ; s'étendre
spreadsheet program	tâbleur
spring	ressort
spruce	épicéa
spy *(v.)*	espionner
square	carré
squash *(n. US, v.)*	courge ; écraser
stack	tas, pile
staff	personnel
stainless steel	acier inoxydable, Inox
staircase	escalier
stapler	agrafeuse
star drive screw	vis à six pans creux
starting switch	commande marche-arrêt
state *(v.)*	déclarer, dire
state-of the-art	de pointe
statement	déposition, déclaration
stator	stator
steel	acier
steel plate	plaque d'acier
steel square	équerre en acier
stocked	stocké
stopover	escale
storage	stockage
storage capacity	capacité de stockage
storage device	périphérique de stockage
store *(v.)*	stocker
stores	provisions, réserves, stocks
straight	droit
strain	charge
strength	force
stretch *(v.)*	s'étendre
strip	bande, bandeau
stripping tongs	pince à dénuder
structural calculation	calcul structural
structural form	forme structurale
stuffed	farci
sturdy	solide
subdivide *(v.)*	subdiviser
subdivision	subdivision
subject	sujet
subject line	sujet, référence
subject to	assujetti à
submit an application	poser sa candidature
substance	substance
substitute	substituer
suburban	de banlieue
success	succès, réussite
successful	réussi
suction	succion, aspiration
sufficient	suffisant
suggestion	suggestion, proposition
suit *(v.)*	convenir à
suitable	approprié, qui convient
sulphur	soufre
sulphuric acid	acide sulfurique
summary	sommaire
sundries	articles divers, extras
superior	supérieur
supplier	fournisseur
supply *(n., v.)*	fourniture, offre ; fournir
support *(n., v.)*	support ; maintenir, soutenir, supporter
sure thing	bien sûr, naturellement
surely	sûrement
surge	à-coup
surgery	chirurgie
surname	nom de famille
survey	sondage
susceptible to	prédisposé à
suspended floor	plancher suspendu
sustained economic growth	croissance économique soutenue
switch	interrupteur
switch on *(v.)*	allumer
synthetics	synthétiques
system disk	disque système

T

table	table, schéma
table hand feed	table à alimentation manuelle
table traverse	table transversale
tacker	cloueuse
tailstock	contre-poupée
take into account *(v.)*	prendre en compte
take into consideration *(v.)*	prendre en considération
take orders	prendre des commandes

tap	taper, tapoter
tape measure	mètre
task	tâche
taxi rank	station de taxis
technical drawing	dessin technique
technology	technologie
tedious	ennuyeux, fastidieux
teeth *(pl.)*	dents
telephone directory	annuaire
tension	tension
term	condition, terme
terms of delivery	conditions de livraison
terms of payment	conditions de paiement
Thanksgiving Day *(US)*	fête nationale qui commémore l'installation des premiers colons (4e jeudi de novembre)
the file is corrupted	le fichier est altéré
the same holds if …	il en est de même si …
theft	vol
thereby	ainsi
therefore	donc
thermoplastic	thermoplastique
thermosetting (plastics)	(plastiques) thermodurcissables
thickness	épaisseur
thorough	minutieux, approfondi
thread	pas, filetage
thrive *(v.)*	réussir, se développer
thus	ainsi, donc
ticket agent	agence de spectacles / de voyages
tidal power	énergie marémotrice
tide	marée
tie *(v.)*	lier
tighten *(v.)*	serrer
tile	tuile
timber	bois de construction
time consuming	qui prend beaucoup de temps
timeless	intemporel, hors du temps
tip	bout, pointe ; pourboire
to and fro	mouvement de va-et-vient
tool kit	jeu d'outils
toolbox	boîte à outils
toolpost	chariot porte-outils
tools	outils
top *(v.)*	napper (sauce)
topple *(v.)*	tomber, s'écrouler, chavirer
torch	lampe
torque	moment de torsion
tough	dur
town hall	mairie
toy	jouet
trade	commerce
trade guild	organisme professionnel
trade union	syndicat
traffic jam	bouchon
train *(v.)*	former
transform *(v.)*	transformer
transformer	transformateur
transmit *(v.)*	transmettre
transparency	transparent
transship *(v.)*	transborder
travel agent / agency	agence de voyages
travel arrangements	organisation de / préparatifs pour le voyage
traveller	voyageur
tray	plateau, casier, corbeille
tremendous	énorme
tremendously	énormément
trench	tranchée
triangle	triangle
triangular	triangulaire
trigger switch	gâchette
troubleshooting	dépannage
trout	truite
troy weight	troy
tube	tube
tungsten	tungstène
tungsten-carbide	carbure de tungstène
turnover	chiffre d'affaires
turpentine	térébenthine
twist drill	perceuse à spirale

U

undo *(v.)*	défaire
uninstall *(v.)*	désinstaller
unit	unité
unlicensed	sans permis, non autorisé
unpredictable	imprévisible
up-to-date	moderne
up-to-dateness	modernité
update *(v.)*	mettre à jour
urban	urbain
urgent	urgent
usually	d'habitude
utility	utilité
utmost	le plus grand

V

vacation *(US)*	vacances
vacuum tube	tube électronique / à vide

valet service	pressing (de l'hôtel)
value	valeur
valve	vanne
variable	variable
varied	varié
various	divers
veal	veau
vegetable	légume
vegetarian	végétarien
venison	venaison
ventilation	ventilation
vernier caliper	jauge micrométrique
versatile	polyvalent
vertical milling machine	fraiseuse verticale
vertical slide	coulisse verticale
vessel	vaisseau
Veterans Day *(US)*	le 11 novembre
via	via, par
vice	étau
vice(-)versa	vice versa
view	vue
virus protector	dispositif de protection contre les virus
visual	visuel
vital	essentiel
voltage (V)	tension
voltage tester	testeur de tension
volume	volume

W

waiter	serveur
waitress	serveuse
wall socket	prise murale
walnut	noix
walnut	noix
walnut	noix, noyer
warehouse	entrepôt
warranty	garantie
waste	déchets
water vapour	vapeur d'eau
wave power	énergie des vagues
wavelength	longueur d'ondes
weak	faible
weaken *(v.)*	affaiblir
wear	usage
weigh *(v.)*	peser
weight	poids
weld *(v.)*	souder
welding	soudure
welding torch	chalumeau
well	puits
well-done	bien cuit (steak)
wheel	roue
wherever	n'importe où
whitewood	bois blanc
wide area network (WAN)	WAN
width	largeur
windowpane	vitre
wire stripper	pince à dénuder
with respect to	en ce qui concerne
witness *(v.)*	témoigner de
wood	bois
word processor	machine de traitement de texte
work *(v.)*	travailler
work permit	permis de travail
workbench	banc de travail, établi
working range	zone de travail
workpiece	pièce
Would you mind copying the faxes for me?	Pourriez-vous me copier les télécopies, s'il vous plaît ? (= Est-ce que cela vous dérangerait de … ?)
write protection	protection en écriture

Y

yard	cour, dépôt
yellow pages	pages jaunes
Yours faithfully	salutations distinguées

Z

zero emission vehicle	voiture à émission(s) zéro
zinc	zinc

Table 1

Table 1: Periodic Table of the Elements

Element	Symbol	Atomic No.	Atomic Wt.	Density	Melting Point °C	French
alumin(i)um (BE)	Al	13	26.9815	2.699	660 °C	aluminium
argon	Ar	18	39.948	0.00178	-189°C	argon
arsenic	As	33	74.9216	1.97; 5.73		arsenic
beryllium	Be	4	9.0122	1.848	1285°C	béryllium
bromine	Br	35	79.904	3.12	-7°C	brome
cadmium	Cd	48	112.41	8.648	321°C	cadmium
calcium	Ca	20	40.08	1.55	851°C	calcium
carbon	C	6	12.011	1.88-3.53	>3550°C	carbone
cesium	Cs	55	132.9054	1.873	29°C	césium
chlorine	Cl	17	35.453	0.00321	-103°C	chlore
chromium	Cr	24	51.996	7.19	1890°C	chrome
cobalt	Co	27	58.9332	8.9	1495°C	cobalt
copper	Cu	29	63.546	8.96	1084°C	cuivre
fluorine	F	9	18.9984	0.00169	-220°C	fluor
gold	Au	79	196.9665	18.88	1063°C	or
helium	He	2	4.0026	0.000179	-271°C	hélium
hydrogen	H	1	1.00794	0.000089	-259°C	hydrogène
iodine	I	53	126.9045	4.93	113°C	iode
iridium	Ir	77	192.22	22.42	2443°C	iridium
iron	Fe	26	55.847	7.874	1100...1600°C	fer
krypton	Kr	36	83.80	0.00373	-157°C	krypton
lead	Pb	82	207.2	11.35	327°C	plomb
lithium	Li	3	6.941	0.534	181°C	lithium
magnesium	Mg	12	24.305	1.738	657°C	magnésium
manganese	Mn	25	54.9380	7.2-7.4	1244°C	manganèse
mercury	Hg	80	200.59	13.546	-39°C	mercure
molybdenum	Mo	42	95.94	10.22	2620°C	molybdène
neon	Ne	10	20.179	0.0009	-249°C	néon
nickel	Ni	28	58.69	8.902	1453°C	nickel
nitrogen	N	7	14.0067	0.00125	-210°C	nitrogène
oxygen	O	8	15.9994	0.001429	-219°C	oxygène
phosphorus	P	15	30.9738	1.8-2.7	44°C	phosphore
platinum	Pt	78	195.08	21.45	1774°C	platine
plutonium	Pu	94	(244)	19.84	640°C	plutonium
potassium	K	19	39.0983	0.862	64°C	potassium
radium	Ra	88	226.0254	5	700°C	radium
radon	Rn	86	(222)	0.00973	-71°C	radon
selenium	Se	34	78.96	4.3; 4.8	221°C	sélénium
silicon	Si	14	28.0855	2.33	1410°C	silicium
silver	Ag	47	107.8682	10.50	961°C	argent
sodium	Na	11	22.9898	0.9712	98°C	sodium
strontium	Sr	38	87.62	2.54	771°C	strontium
sulf(ph)ur (BE)	S	16	32.06	1.96-2.07	120°C	soufre
tin	Sn	50	118.69	5.75-7.31	232°C	étain
titanium	Ti	22	47.88	4.54	1677°C	titane
tungsten	W	74	183.85	19.3	3380°C	tungstène
uranium	U	92	238.0289	18.95	1132°C	uranium
vanadium	V	23	50.9415	6.11	1919°C	vanadium
xenon	Xe	54	131.29	0.00589	-112°C	xénon
zinc	Zn	30	65.38	7.133	420°C	zinc

Table 2: Conversion table (Celsius – Fahrenheit)

Formula: °F = °C x 9 / 5 + 32 / °C = (°F - 32) x 5 / 9

°C	°F	°C	°F	°C	°F
-20	-4	30	86	41.4	106.52
-19	-2.2	31	87.8	41.5	106.7
-18	-0.4	32	89.6	41.6	106.88
-17	1.4	33	91.4	41.7	107.06
-16	3.2	34	93.2	41.8	107.24
-15	5	35	95	41.9	107.42
-14	6.8	36	96.8	42	107.6
-13	8.6	37	98.6	43	109.4
-12	10.4	37.1	98.78	44	111.2
-11	12.2	37.2	98.96	45	113
-10	14	37.3	99.14	46	114.8
-9	15.8	37.4	99.32	47	116.6
-8	17.6	37.5	99.5	48	118.4
-7	19.4	37.6	99.68	49	120.2
-6	21.2	37.7	99.86	50	122
-5	23	37.8	100.04	55	131
-4	24.8	37.9	100.22	60	140
-3	26.6	38	100.4	70	158
-2	28.4	38.1	100.58	80	176
-1	30.2	38.2	100.76	90	194
0	32	38.3	100.94	100	212
1	33.8	38.4	101.12	110	230
2	35.6	38.5	101.3	150	302
3	37.4	38.6	101.48	200	392
4	39.2	38.7	101.66	300	572
5	41	38.8	101.84	400	752
6	42.8	38.9	102.02	500	932
7	44.6	39	102.2	600	1112
8	46.4	39.1	102.38	700	1292
9	48.2	39.2	102.56	800	1472
10	50	39.3	102.74	900	1652
11	51.8	39.4	102.92	1000	1832
12	53.6	39.5	103.1		
13	55.4	39.6	103.28		
14	57.2	39.7	103.46		
15	59	39.8	103.64		
16	60.8	40	104		
17	62.6	40.1	104.18		
18	64.4	40.2	104.36		
19	66.2	40.3	104.54		
20	68	40.4	104.72		
21	69.8	40.5	104.9		
22	71.6	40.6	105.08		
23	73.4	40.7	105.26		
24	75.2	40.8	105.44		
25	77	40.9	105.62		
26	78.8	41	105.8		
27	80.6	41.1	105.98		
28	82.4	41.2	106.16		
29	84.2	41.3	106.34		

°F	°C	°F	°C	°F	°C
-4	-20.00	46	7.78	96	35.56
-3	-19.44	47	8.33	97	36.11
-2	-18.89	48	8.89	98	36.67
-1	-18.33	49	9.44	99	37.22
0	-17.78	50	10.00	100	37.78
1	-17.22	51	10.56	101	38.33
2	-16.67	52	11.11	102	38.89
3	-16.11	53	11.67	103	39.44
4	-15.56	54	12.22	104	40.00
5	-15.00	55	12.78	105	40.56
6	-14.44	56	13.33	106	41.11
7	-13.89	57	13.89	107	41.67
8	-13.33	58	14.44	108	42.22
9	-12.78	59	15.00	109	42.78
10	-12.22	60	15.56	110	43.33
11	-11.67	61	16.11	200	93.33
12	-11.11	62	16.67	250	121.11
13	-10.56	63	17.22	300	148.89
14	-10.00	64	17.78	400	204.44
15	-9.44	65	18.33	500	260.00
16	-8.89	66	18.89	600	315.56
17	-8.33	67	19.44	700	371.11
18	-7.78	68	20.00	800	426.67
19	-7.22	69	20.56	900	482.22
20	-6.67	70	21.11	1000	537.78
21	-6.11	71	21.67	2000	1093.33
22	-5.56	72	22.22	3000	1648.89
23	-5.00	73	22.78	4000	2204.44
24	-4.44	74	23.33	5000	2760.00
25	-3.89	75	23.89	6000	3315.56
26	-3.33	76	24.44	7000	3871.11
27	-2.78	77	25.00	8000	4426.67
28	-2.22	78	25.56		
29	-1.67	79	26.11		
30	-1.11	80	26.67		
31	-0.56	81	27.22		
32	0.00	82	27.78		
33	0.56	83	28.33		
34	1.11	84	28.89		
35	1.67	85	29.44		
36	2.22	86	30.00		
37	2.78	87	30.56		
38	3.33	88	31.11		
39	3.89	89	31.67		
40	4.44	90	32.22		
41	5.00	91	32.78		
42	5.56	92	33.33		
43	6.11	93	33.89		
44	6.67	94	34.44		
45	7.22	95	35.00		

Mathematical terms and symbols

Addition: 2+3 = 5 (Two **plus** 3 **equals** five.)

Subtraction: 8-4 = 4 (Eight **minus** 4 **equals** four)

Multiplication: 9x4 = 36 (Nine **multiplied by** four **equals** thirty-six; or nine **times** four …)

Division: 55÷11 = 5 (Fifty-five **divided by** eleven **equals** five.)

$\sqrt{}^2$ – square root: $9\sqrt{}^2$ the square root of 9 equals three

$\sqrt{}^3$ – cube root $8\sqrt{}^3$ the cube root of 8 is 2

3^2 – three squared equals 9

4^3 – four cubed equals 64

2^4 – two to the power of 4 equals 16

Fractions

$^1/_2$ – one half, a half

$^1/_3$ – one third, a third

$^1/_4$ – one quarter, a quarter

$^3/_4$ – three quarters

$^1/_5$ – one fifth, a fith

$^5/_8$ – five eighths

% per cent

‰ per thousand

≠ does not equal, is not equal to

> is more than / greater than

< is less than

° degree

Conversion table inch / foot – cm / m

1 inch (in) = 2.54 cm
1 foot = 30.48 cm
(approximate conversion)

Examples:

- A length of 35 cm would be about 13 $^3/_4$" or 1' 2" $^3/_4$" (1 foot 2 $^3/_4$ inches).
- A person who is 1,79 m tall would be 5' 10" (five feet, ten inches) often abbreviated idiomatically to: five foot ten (note the singular: foot).
- Feet are used internationally to measure the distance of aircraft from the ground, e.g.10,000 ft.

in	cm	cm	in
1	2.54	3	1.18
2	5.08	4	1.57
3	7.62	5	1.97
4	10.16	10	3.94
5	12.70	15	5.91
6	15.24	20	7.87
7	17.78	25	9.84
8	20.32	30	11.81
9	22.86	35	13.78
10	25.40	40	15.75
11	27.94	45	17.72
12	30.48	50	19.69

ft	m	m	ft
1	0.3048	1	3.28
2	0.6096	2	6.56
3	0.9144	3	9.84
4	1.2192	4	13.12
5	1.5240	5	16.40
6	1.8288	6	19.68
7	2.1336	7	22.96
8	2.4384	8	26.24
9	2.7432	9	29.52
10	3.0480	10	32.80
11	3.3528	11	36.08
12	3.6576	12	39.36

Conversion table yard – metre / mile – kilometre (approximate conversions)

1 yard (yd) = 0.9144 metres (m) **1 mile = 1.609 km**

yd	m	m	yd	mile	km	km	mile
1	0.914	1	1.09	1	1.609	1	0.62
2	1.829	2	2.19	2	3.218	2	1.24
3	2.743	3	3.28	3	4.827	3	1.86
4	3.658	4	4.37	4	6.436	4	2.49
5	4.572	5	5.47	5	8.045	5	3.11
6	5.486	6	6.56	6	9.654	6	3.73
7	6.401	7	7.66	7	11.263	7	4.35
8	7.315	8	8.75	8	12.872	8	4.97
9	8.230	9	9.84	9	14.481	9	5.59
10	9.144	10	10.94	10	16.090	10	6.22
11	10.058	11	12.03	11	17.699	11	6.84
12	10.973	12	13.12	12	19.308	12	7.46
13	11.887	13	14.22	13	20.917	13	8.08
14	12.802	14	15.31	14	22.526	14	8.70
15	13.716	15	16.40	15	24.135	15	9.32
16	14.630	16	17.50	16	25.744	16	9.94
17	15.545	17	18.59	17	27.353	17	10.57
18	16.459	18	19.69	18	28.962	18	11.19
19	17.374	19	20.78	19	30.571	19	11.81
20	18.288	20	21.87	20	32.180	20	12.43
25	22.860	25	27.34	25	40.225	25	15.54
30	27.432	30	32.81	30	48.270	30	18.65
35	32.004	35	38.28	35	56.315	35	21.75
40	36.576	40	43.74	40	64.360	40	24.86
45	41.148	45	49.21	45	72.405	45	27.97
50	45.720	50	54.68	50	80.450	50	31.08
55	50.292	55	60.15	55	88.495	55	34.18
60	54.864	60	65.62	60	96.540	60	37.29
70	64.008	70	76.55	70	112.630	70	43.51
80	73.152	80	87.49	80	128.720	80	49.72
90	82.296	90	98.43	90	144.810	90	55.94
100	91.440	100	109.36	100	160.900	100	62.15
150	137.160	150	164.04	150	241.350	150	93.23
200	182.880	200	218.72	200	321.800	200	124.30
250	228.600	250	273.40	250	402.250	250	155.38
300	274.320	300	328.08	300	482.700	300	186.45
350	320.040	350	382.76	350	563.150	350	217.53
400	365.760	400	437.45	400	643.600	400	248.60
450	411.480	450	492.13	450	724.050	450	279.68
500	457.200	500	546.81	500	804.500	500	310.75
600	548.640	600	656.17	600	965.400	600	372.90
700	640.080	700	765.53	700	1126.300	700	435.05
800	731.520	800	874.89	800	1287.200	800	497.20
900	822.960	900	984.25	900	1448.100	900	559.35
1000	914.400	1000	1093.61	1000	1609.000	1000	621.50

Acknowledgements

page 7 Klett-Archiv
page 10 Leeds Bradford International Airport
page 11 Leeds Bradford International Airport
page 12 Elmar Feuerbach, Remshalden
page 19 Topdeq GmbH
page 20 Elmar Feuerbach, Remshalden; Xerox GmbH, Düsseldorf
page 23 Elmar Feuerbach, Remshalden
page 33 IBM Deutschland; Elmar Feuerbach, Remshalden
page 34 IBM Deutschland
page 35 Elmar Feuerbach, Remshalden
page 37 (h): Elmar Feuerbach, Remshalden; (b): IBM Deutschland
page 41 Elmar Feuerbach, Remshalden
page 43 (h): 3-D view, AGS GmbH, Leonberg; (b): Reis GmbH & Co Maschinenenfabrik Obernburg
page 49 Graphikbüro Böttcher & Bayer, Stuttgart
page 50 MEV (AFA 29)
page 55 Siemens – Pressebild
page 57 EMCO Maier Ges. m.b.H. (A-5400 Hallein-Taxach), Clarke International, Hahn & Kolb, Elmar Feuerbach, Remshalden
page 62 EMCO Maier Ges. m.b.H. (A-5400 Hallein-Taxach)
page 63 Brinkmann + Wecker GmbH, General-Importeur + Service, Paderborn
page 64 EMCO Maier Ges. m.b.H. (A-5400 Hallein-Taxach)
page 65 EMCO Maier Ges. m.b.H. (A-5400 Hallein-Taxach)
page 66 Elmar Feuerbach, Remshalden
page 67 Clarke International
page 69 Clarke International
page 70 Elmar Feuerbach, Remshalden
page 73 Hahn & Kolb; Elmar Feuerbach, Remshalden
page 76 Clarke International
page 77 Elmar Feuerbach, Remshalden
page 78 Elmar Feuerbach, Remshalden
page 81 Elmar Feuerbach, Remshalden
page 85 Kühlschrank, Geschirrspülmaschine: Bauknecht Hausgeräte GmbH, Schorndorf; Staubsauger: © The Hoover Company 2001 (a division of MAYTAG); Handy: Elmar Feuerbach, Remshalden; TV: SONY
page 87 (hg): Elmar Feuerbach, Remshalden; (b): Graphikbüro Böttcher & Bayer, Stuttgart
page 88 (h): Bosch; (b): Graphikbüro Böttcher & Bayer, Stuttgart
page 89 Elmar Feuerbach, Remshalden
page 90 MEV (FCC 1)
page 91 Elmar Feuerbach, Remshalden
page 92 Bauknecht Hausgeräte GmbH, Schorndorf
page 95 Graphikbüro Böttcher & Bayer, Stuttgart
page 96 (h): Elmar Feuerbach, Remshalden; dessin: Graphikbüro Böttcher & Bayer, Stuttgart
page 97 Daimler-Chrysler
page 98 Graphikbüro Böttcher & Bayer, Stuttgart
page 99 Elmar Feuerbach, Remshalden
page 100 Graphikbüro Böttcher & Bayer, Stuttgart
page 101 Ford-Werke AG
page 102 Rolf Benz AG, Nagold
page 103 Elmar Feuerbach, Remshalden
page 105 HANS BARRO Aluminium-Bootsbau, D-89293 Kellmuenz
page 106 (h): © COREL; (b): Elmar Feuerbach, Remshalden
page 107 KYOCERA Fineceramics GmbH
page 108 Elmar Feuerbach, Remshalden; (milieu): TEFAL Deutschland GmbH
page 111 Elmar Feuerbach, Remshalden
page 112 Elmar Feuerbach, Remshalden
page 120 Klett-Archiv
page 121 Siemens – Pressebild
page 122 (h): Siemens-Pressebild; (b): MEV
page 123 © PhotoDisc
page 124 (h): MEV (AFA 22); (b): Klett-Archiv © COREL
page 125 © CORBIS/Royalty-Free
page 126 (h) MEV (AFA 22) ; Corbis/Roger Ressmeyer
page 130 Elmar Feuerbach, Remshalden
page 131 Dr. Wolfgang Schäfer, Buchen
page 137 L. Rofe
page 140 MEV (AFA 26)
page 141 © Terragraphix Inc.
Seite 142 Dr. Wolfgang Schäfer, Buchen
page 143 © COREL
page 145 © PhotoDisc
page 147 Klett-Archiv
page 148 (h): SIEMENS AG, München; (b): Elmar Feuerbach, Remshalden
page 152 © CORBIS/Royalty-Free
page 157 Klett-Archiv
page 161 (g). MEV (AFA 18); (d). MEV (AFA 30)
page 162 (g). MEV (AFA 20); (d). MEV (AFA 25)
page 164 (g). MEV (AFA 31); (d). MEV (AFA 30)
page 165 MEV (AFA 27)
page 166 Elmar Feuerbach, Remshalden
page 167 Elmar Feuerbach, Remshalden
page 168 (h): Elmar Feuerbach, Remshalden; (b): Cosmos / Westlight / L. Hawring
page 169 (h): © PhotoDisc; (b): © CORBIS/Royalty-Free
page 171 MEV (AFA 19)
page 172 (hg et bd): © CORBIS/Royalty-Free; (hd et mg) : MEV (AFA 19); M.r. MEV (FAC 13/14); (g): © PhotoDisc

Imprimé en France par Clerc à Saint-Amand-Montrond
N° d'imprimeur : 8804 - N° d'édition : 003882-01
Dépôt légal : août 2004